Data Algebra

JabrDB and its Relatives

by Robin Bloor, Ph.D.

JDL

Jabr Data Labs

Data Algebra
JabrDB and its Relatives

First Edition: July 2026

Library of Congress Control Number: 2026916765

ISBN 978-1-957278-19-3

Printed in the United States of America

Jabr Data Labs
Cedar Parlk, Texas

Dedication

To Professor Gary Sherman, the father of Data Algebra, and the grandfather of JabrDB

"The time has come," the Walrus said,
"To talk of many things,
Of shoes and ships and sealing wax,
and cabbages and kings,
And if the sea is boiling hot,
and whether pigs have wings."

~ Lewis Carroll

Data Algebra

JabrDB, and its Relatives

Table of Contents

Introduction 1
Chapter 1: What Is? 3
Chapter 2: In Search of a Data Algebra 11
Chapter 3: Cantor, Zermelo and Fraenkel 23
Chapter 4: Descartes' Legacy 43
Chapter 5: The Lion Sleeps Tonight 57
Chapter 6: Yin and Yang 71
Chapter 7: The Gathering of the Clans 91
Chapter 8: Joinery 121
Chapter 9: Unleashing The Hordes 149
Chapter 10: Modest Math Built Our World 155
Chapter 11: Good Math, Bad Math 165
Chapter 12: JabrDB, a Reunion of Broken Parts 175
Chapter 13: The Sorry Saga of SQL 191
Chapter 14: JQL v SQL 201
Chapter 15: The Event Horizon 225
Appendices 233

Introduction

Mathematics is a more powerful instrument of knowledge than any other that has been bequeathed to us by human agency.

~ Descartes

There is a peculiarity at the heart of information technology that few who work in it ever pause to notice: the field has never possessed a proper mathematics of its own data.

Essentially, computers do one thing — they transform data into other data — and yet the data itself, the very substance the whole enterprise exists to handle, does not rest on a rigorous foundation. It rests only on convention, a convention that has evolved unexamined for over 70 years.

This was not inevitable. Efforts were made in the early 1970s, born of a genuine mathematical ambition; the relational model of data was meant to put data on a sound footing. Unfortunately it was not created and pursued by skilled mathematicians. It succeeded only partially. There were errors.

This book is devoted to finishing the work that was begun then, but set aside. Its subject is the algebra of data: a single mathematical structure, built in four levels — couplets, relations, clans and hordes — into which all data, of every kind and in every context, can be placed, and through which it can be manipulated with precision. The algebraic structure is closed and complete. Every operation a database or a file system performs can be expressed within it, composed from it, and reasoned about in it. No shape of data — table, document, graph, key-value pair, hierarchy — falls outside its reach.

This algebra was invented by the late Gary Sherman, a mathematics professor whose rigor runs through every level of it. When it first appeared over 15 years ago, there were efforts to build database optimization technology with it. Technically, it succeeded, but sadly it never saw commercial success and interest in the algebra behind it waned.

But it never died. Over the past couple of years, data algebra has been put to work to build JabrDB, a native algebraic database. It is real, it exists and if you care to you can install it. You can load your own data, and watch the algebra do its thing. Around this core capability, a family of related applications and a connectivity architecture is being built.

Visit JabrDataLabs.com if you want to download it.

Our intention is that this algebra should eventually become the foundation of all data technology, eventually reaching out to and touching the billions of data stores that make up the universe of data.

Chapter 1: What Is?

There is nothing more difficult to take in hand, more perilous to conduct, or more uncertain in its success, than to take the lead in the introduction of a new order of things.

~ Machiavelli

WE TOOK Machiavelli's advice to heart. The algebra of data is a new application of mathematics. It has existed as a set of mature usable techniques since 2012. The few experienced mathematicians who are currently familiar with it will tell you there is nothing new about it, in the pure mathematical sense. For example, there was no need to adjust or extend set theory to formulate the various algebras that constitute the algebra of data – yes, indeed, data algebra comprises more than one algebra.

It applies set theory in a new way, and for that reason, it constitutes a new branch of applied mathematics. It was first formulated as a collection of mathematical constructs that were thought to provide a practical foundation for data management software. It was gradually tested and applied to data management problems, first to "relational" data structures and then to "graph" data structures.

This initial chapter is devoted to answering several questions that all begin with the words "What is..." We believe they are worth discussing before we introduce the algebra of data. The first such question is:

What is mathematics?

The *New Oxford American Dictionary* defines it as:

> The abstract science of number, quantity, and space. Mathematics may be studied in its own right (pure mathematics), or as it is applied to other disciplines such as physics and engineering (applied mathematics).

Alternatively, the *Merriam-Webster* dictionary says:

> The science of numbers and their operations, interrelations, combinations, generalizations, and abstractions and of space configurations and their structure, measurement, transformations, and generalizations.

In fact, neither of these definitions is entirely adequate. The first confines mathematics to number, quantity and space, when in practice, some areas of mathematics deal with "things" that may not be spatial or numeric or have the attribute of quantity. Data can be such a thing.

The second definition is slightly better, as it refers directly to operations, interrelations, combinations, generalizations and abstractions, which are the beating heart of mathematics. Nevertheless, this definition also confines mathematics to numeric and spatial concerns.

The power of mathematics lies partly in its ability to construct useful models of real things. Using scale measurements, geometry and various calculations, you can design buildings. With more sophisticated mathematics, you can design chemical plants or airplanes. Mathematics is both the toolbox of innovators and the workhorse of engineers.

Even where mathematics has failed to model a domain perfectly – for example, weather prediction – it has no peer. Weather forecasts still rely on mathematical models, and nowadays, given sufficient computer power and historical data, they are usually accurate for two days or more.

In fact – and this is important to understand – mathematics is never wrong per se. As Isaac Newton remarked, "Errors are not in the art but in the artificers." A mathematical statement can be incorrect if someone makes an error in formulating it. A mathematical model can be horribly inaccurate if the modeling work is done poorly. But the mathematics itself is blameless. The tool is either used well or badly. This can be said of very few areas of study.

Data algebra models data, and it does so with rigorous accuracy. This attention to detail makes it an extremely useful tool for anyone who wants to define data structures and manipulate them algebraically.

The next question is also a fundamental one:

What is a computer?

Dictionaries focus on the fact that a computer is an electronic machine that runs programs to process information. However, here we need to be more particular and define a computer in terms that have relevance to data algebra.

So think of a computer as an electronic machine that runs programs that *act on data*. This brief, accurate definition allows us to make an important point:

> *The only thing a computer ever does is take data and transform it into other data.*

Every program that you have ever used or can conceive of does this and only this. It is true of operating systems, spreadsheets, email and big databases. It is true of every executable program.

There is an extremely powerful corollary to this: given that a data algebra exists, ALL programs can be written in algebraic terms as a set of mathematical functions that transform various algebraic collections of data into other algebraic collections of data.

Whether this means that ALL programs should be written in a strict algebraic manner is a moot point. In reality, simple programs will gain little from it.

What is an algebra?

Back to the dictionaries. *Merriam-Webster's* answer is:

> Any of various systems or branches of mathematics or logic concerned with the properties and relationships of abstract entities (as complex numbers, matrices, sets, vectors, groups, rings, or fields) manipulated in symbolic form under operations often analogous to those of arithmetic...

This is quite an accurate description of algebra. Algebraic statements use symbols (+, −, ×, ÷, ±, =, ≠, ∈, ∪, and many others) as a shorthand to express specific relationships quickly and succinctly. Such symbols form a kind of lexicon that mathematicians quickly learn. But some people find these symbols difficult to follow and are turned off, perhaps unnecessarily, from mathematics.

To minimize this problem for readers, we have gone to some lengths to translate algebraic statements into words and provided illustrations as well. Of course, those who quickly pick up the symbols can skip these aids.

There are many algebras.

Wikipedia provides a list of over fifty distinct algebras, each focusing on different targets – and it is not an exhaustive list. Data algebra itself is not a single algebra, as we have already noted. You will appreciate what this means as you progress through the book. You will realize that it is, in reality, a number of algebras that are included, one within another, like Russian nesting dolls.

Algebras are specifically built to enable abstraction and generalization. In algebra, the use of alphabetic letters or other symbols to represent real things is a fundamental tactic. So when we say, for example, a locomotive is travelling due west at 60 mph, we may refer to the locomotive as, say, L and its velocity as v. Thus we abstract a real-world object by using a single letter and its velocity by using another letter.

If we manage through various manipulations to find relationships between some of these letters, regardless of their value in any given context, we have a generalization that can be applied to many contexts, which may be very useful. One simple example is the equation you probably learned at school:

$$v(t) = v_0 + at,$$

where v = velocity, v_0 = initial velocity, a = acceleration and t = time.

Discovering such relationships is the primary purpose of any given algebra when it is used to model some real-world situation. However, with data algebra the situation is slightly different. The goal of the algebra is to enable data to be handled algebraically. It does not model the data, it defines it. The rewards that the algebra naturally bestows are generalizations that can be applied to all data. It can solve problems in data integration and software optimization that might otherwise be intractable.

What is data?

In order to construct a data algebra, it is necessary to find an answer to the question "What is data?" and, having found an answer, to discover productive ways to represent data algebraically. Let's satisfy your immediate curiosity – assuming you have some – by discussing the question.

So, if you are willing to play, you are up. Think about the question "What is data?" for a moment or two, and then write your answer in pencil in the space below, without peeking at the text that follows:

Dictionary definitions of data make assertions of the following kind:

- Facts and statistics collected together for reference or analysis.
- Factual information used as a basis for reasoning.
- Any representations to which meaning is or might be assigned.
- The plural of datum.

Thesauri suggest confusing synonyms including: information, aggregation, collection, accumulation, assemblage, fact, details and even input. Perhaps this is what we should expect from the compilers of dictionaries and thesauri, but it's a little disappointing, and it doesn't fry any fish.

The responses to this question from IT experts in the field, often preceded by a long pause and punctuated with some facial contortions and guttural utterings, are more telling. We have experienced the following replies:

- "Knowledge."
- "Facts."
- "Files, digital information."
- "A database."
- "Anything that somebody might want to know that is enumerated, listed or quantified."
- "It is the answer to a question (may be a hypothetical question)."
- "It can be written (this includes that it is limited)."
- "It's data. It is what it is."
- "In computer parlance, it literally means anything that has been captured in any form on any digital recording medium."

We have even met with the somewhat defensive, "What do you mean by that?" and the somewhat facetious, "A character on Star Trek: The Next Generation."

This sample of responses is a hodgepodge of ambiguity and circularity which, if left undisturbed, precludes a rigorous mathematical discussion of data, and since that's what we're focused on, we must do better. If we do not, we will simply repeat the previous errors of the IT industry, which, for its sins, gave birth to SQL, a mathematical disaster of the first order.

To paraphrase Carl Sagan (*Dragons of Eden*), those closest to the intricacies of data seem to have a more highly developed (and ultimately erroneous) sense of its mathematical intractability than those at some remove.

As it turns out, the wife of one of the authors, a notorious data-phobe, provided the most suggestive responses to both questions.

"What is data?" was met with a quizzical look, then, "Well, come on now, you and I both know what it means."

And the subsequent question "What is a datum?" was met with a more menacing look, and "You and I both know what it means!"

Indeed, anybody making a living in the data world knows what data is, and knows what a datum is, and uses the terms with reckless abandon in discussions with colleagues who also know what data is and know what a datum is, even though such "undoubted knowledge" calls into question the very definition of the word "know."

The only implied constant in such discussions is suggested by the surprisingly useful dictionary definition: "data is the plural of datum." That is, data (whatever it is) is a collection of "datums," and a datum (whatever that is) may be collected with other "datums" to create data.

Our assertion here is that the key to providing a practical mathematical understanding of data is to take data and datum as primitive undefined terms and axiomatize their relationship, i.e., the notion of belonging.

Wait! What?

To tell the truth, this was done long ago, in the 19th century by Georg Cantor, as we shall explain in Chapter 3, and the only difference with what we assert and what Georg Cantor asserted is that we are using the terms data and datum, whereas he used the concepts of set and element, respectively.

To put it bluntly, data is to datum as set is to element.

It naturally follows that any mathematically legitimate notion of data algebra must be couched in terms of (and subject to the rigor of) set theory and be independent of any preconceived notions about data: visual, structural or otherwise.

We hope to explain, as simply as we can, the algebraic tale of data. It is a tale of sets recounted by ascending towers of sets of ever increasing cardinality, in the words of Paul Halmos, "of frightening height and complexity."

But don't be too concerned; all algebraic tales are like that.

A final note

This book adopts the style of ending this chapter and each one that follows, with one or more pages entitled Taking Stock. They are provided so that readers can go to these pages to remind themselves of the major points covered in the chapter.

And so...

Taking Stock

The following bullet points summarize this chapter:

- When applied, mathematics is useful by virtue of its ability to model aspects of the real world. The data algebra described in this book accurately models a computer's use of data.
- The only thing computer software ever does is take a collection of data and transform it into another collection of data.
- Algebras are useful because of their abstraction and generalization capabilities. Naturally, this is true of data algebra.
- The question "What is data?" is, on reflection, not such a simple question to answer. Our initial response to this question is to note that the natural habitat of data is set theory.
- But if this is so, how do we account for the apparent fact that real-world data is neither homogeneous nor necessarily algebraic? Indeed, it seems to be an unruly mess compared to one's usual perception of a well-behaved mathematical space. We consider this question in greater depth in Chapter 5.
- This book is complemented by JabrDB, a fully functional algebraic database. It can be downloaded at JabrDataLabs.com.

Chapter 2: In Search of a Data Algebra

We all want progress, but if you're on the wrong road, progress means doing an about-turn and walking back to the right road; in that case, the man who turns back soonest is the most progressive.

~ C. S. Lewis

THE ALGEBRA OF DATA was invented by a mathematics professor who was an early employee of Algebraix Data Corporation, a company that was established to develop software products and services based on data algebra. The founders of Algebraix Data were software engineers.

At the time the company was founded, the notion of a data algebra was only a collective gleam in the founders' eyes. The data algebra that was gradually built and mathematically tested in a variety of data processing contexts took about 5 years to evolve. In parallel with this, the company worked on building data management technology that employed the evolving algebra.

The fact that a data algebra never existed until a handful of years ago will likely surprise you. If you are not a software engineer, you probably assume that such an algebra was developed many decades ago and is regularly employed by software engineers to resolve tricky engineering problems they encounter. Alternatively, you may be a software engineer, in which case you may believe that the universe of data, in its extraordinary variety, is simply too complex and convoluted for mathematics to be usefully employed. Both these perspectives are wrong.

The original concept of what a computer could be was mathematical. It was a concept invented by mathematician, Alan Turing, in 1936. Later, he made important contributions to the design of the electromechanical and electronic code-breaking machines that were built and successfully employed during the Second World War. These machines were, in reality, single purpose computers.

However, in the wake of the Second World War, mathematics played only a subsidiary role in the evolution of computers. Most of the advances in computer hardware and computer software that occurred in those years were achievements of electronic engineering. This unfolded in the following way.

In 1945, mathematician and physicist John von Neumann (and others) wrote a paper entitled ***First Draft of a Report on the EDVAC***. EDVAC was an acronym for Electronic Discrete Variable Automatic Computer. During the war, von Neumann had also been working on difficult mathematical problems, particularly the calculation of artillery tables to better enable the effectiveness of artillery. Electronic calculators had been used to great effect for that purpose.

In the paper, von Neumann described what has come to be called "The von Neumann computer architecture." This pragmatic architecture involved a processing unit embodying an arithmetic logic unit and processor registers, a control unit containing an instruction register and program counter and memory to store both data and instructions. It also included external mass storage and input and output mechanisms for data. The model, rather than an architecture based on a Turing Machine, established itself at the foundation of the computer industry that gradually emerged.

Mathematicians had some involvement in the further evolution of computers. But there were few efforts to model the processing of data in a mathematical fashion, with data conforming to elements or collections of elements within an algebraic domain and program instructions conforming to algebraic operations on that data – and those efforts were unsuccessful.

Shaving the barber

The barber problem is also known as Russell's paradox, since the famed philosopher and mathematician, Bertrand Russell, popularized it. He used it as a means of pointing out a fundamental problem in Georg Cantor's definition of set theory. The barber problem is this:

> A regiment in the army has a designated barber. The colonel in control of the regiment gives orders to the barber. He commands thus, "From today, every man in the regiment shall be clean-shaven. Henceforth you must shave every man who does not shave himself."

The problem is: who shaves the barber?

Logically, the barber cannot shave himself since he is commanded to shave men who do not shave themselves. And yet if he does not shave himself, he becomes one of the men whom he should shave.

The barber problem is a variation of a familiar Greek paradox (the circular "All Cretans are liars. I am a Cretan"). Its relevance to set theory is that when set theory was originally created by Georg Cantor, a set was deemed to be any definable collection. Russell demonstrated that sets needed to be defined more precisely than that, by considering those sets that are "members of themselves." The problem, which is identical to the barber problem, can be described mathematically as follows:

> Let *R* be the set of all sets that are not members of themselves. If *R* is not a member of itself then by definition it must contain itself. But if it contains itself then it violates its own definition as the set of all sets that are not members of themselves. We can express this algebraically as follows:
>
> $$\text{Let } R = \{A\colon A \notin A\}, \text{ then } R \in R \text{ if, and only if, } R \notin R,$$
>
> i.e., let *R* equal the sets *A*, such that *A* is not an element of *A*, then "*R* is an element of itself" implies "*R* is not an element of itself" and vice versa.

Mathematically, the consequence of the barber problem was that the axioms of set theory were reviewed and changed so that such paradoxes were eliminated. Subsequently, set theory became a foundational branch of mathematics, partly because most mathematical objects could be represented within set theory. All data, whatever form it takes, can also be represented as a set or an element of a set.

Functional programming

Despite the computer's evolutionary path, computational mathematics did not die simply because the von Neumann architecture became dominant. In fact, a good deal of work was done on what came to be called lambda calculus (λ-calculus), which in turn played an important role in the theory of programming languages.

We will not describe lambda calculus here, but nevertheless we feel obliged to draw attention to the Church-Turing Thesis, which states that "the untyped lambda calculus is capable of computing all effectively calculable functions." Put simply, it means that mathematicians have a computational model that they believe to be complete.

A natural consequence of the invention of lambda calculus was its use in development of various programming languages. Such languages are generally referred to as functional programming languages, and they resemble algebraic activity by virtue of their applying functions to carry out calculations and transformations. In other words, functional programming languages process data in a mathematical way. What they are missing is a data algebra to define the data.

The tower of Babel

Almost everyone who enters the software world quickly becomes astonished at the number of different programming languages that exist. It has even been suggested that there are now more programming languages than there are human languages. For the record, there are about 6000 human languages, but only about 100 of those languages are spoken by significant populations of people.

The same could be said of programming languages. There are certainly thousands. Wikipedia lists 600 or so languages that it regards as notable. If you read through the list you may encounter many that you have never heard of before, and yet there are many, many more, thousands more, that never made the cut.

Programming languages proliferate for a variety of reasons. Often innovations in computer hardware provoke the creation of new languages. Sometimes the need to specialize a language for a particular area of application (statistics, graphics, video rendering, text processing, etc.) forges a new language. Sometimes software companies wish to create a proprietary language they can control. Sometimes it's simply a matter of evolution. New problems are encountered, new features are added, and thus new dialects are born. Some languages (notably object-oriented ones) are extensible, and hence they evolve of their own accord.

The evolution of functional languages

The first language with functional features was Lisp, developed by John McCarthy (at MIT) in the late 1950s. Other languages such as IPL and APL followed in its wake. In 1977, John Backus formally introduced the idea of functional programming in a Turing Award lecture entitled: *Can Programming Be Liberated From the von Neumann Style? A Functional Style and its Algebra of Programs.* He defined functional programs as being constructed in a hierarchical way by means of "combining forms" that allow an "algebra of programs." His paper generated enough interest to provoke research into functional programming.

This gave rise to a whole raft of new programming languages, including: ML, SASL, Miranda, NPL, Hope, J, K and Q. Then, in 1987, there was a general move to form an open standard for functional programming, and the Haskell language was born. Haskell, an open source language, has been evolving ever since then.

Functional programming languages work in a particular way. They are ***declarative.*** This means that the program tells the computer what to do without telling it how to do it. This is in contrast to the procedural or imperative style of programming, where the language specifically controls the flow (the procedure) of the computer's activity.

Declarative is an umbrella term in the sense that it covers a number of different language behaviors. In the case of functional programming languages, the language implementation tries to ensure that there are no side effects to the execution of any functions (bugs in procedural programs often turn out to be unexpected side effects of the program code).

So a functional program is a collection of functions that will be executed as specified by the programmer. Ideally, all state changes to any collections of data will be represented as functions. All of this puts an onus on the compiler to resolve how the computer will execute the program. However, it should make it much easier for a programmer to understand and predict the behavior of a program he or she writes.

Hybrid languages

Although the solution of programming problems may in theory be best addressed by thinking of the problem in algebraic terms and formulating a solution by writing algebraic functions, with some applications, it is really quite difficult for programmers to write code in that manner. This has caused a natural objection to a functional programming language.

For example, algebraic functions are natural features of the R language, which focuses on statistical calculations and algorithms, and the Julia language, which focuses on scientific and high performance computing. It is less obvious why you would need such capability in languages like Perl, Javascript, Visual Basic or even the almost defunct COBOL. All are general purpose languages that are more likely to be used for text processing or transactional applications than algebraic manipulations.

Perhaps you can get the best of both worlds with a language that implements function calls after the fashion of a functional programming language but also has procedural features for handling situations where writing procedural code (telling the computer how to do something) is easier for the programmer.

There are, in fact, many such hybrid languages, including Python, Ruby, Scala, Java and Erlang. Among these, Python is the most popular, which is one of the reasons that Algebraix Data Corporation has chosen to use it as the basis of its Algebraix language.

And Yet, No Data Algebra...

Our brief foray into the evolution of functional programming languages shows that there have been pioneers who tried to exploit mathematical ideas in software. It is well known that the dominance of the von Neumann architecture led to a less mathematically oriented computer industry, but what the practical consequences were and what to do about it has never been clear.

The enthusiasm for functional programming was and is an enthusiasm for mathematics. It was fostered by individuals who were convinced that there were significant advantages to a mathematical style of programming. And there is virtue in it. When such a programming style is enshrined in a programming language, and competently employed, it gives rise to fewer programming errors, requires fewer lines of code and promotes software reuse. In short, it can improve productivity significantly.

Nevertheless, while there have been efforts to mimic mathematics in software, there has been no successful attempt to define an explicit algebra of data that enables data to be dealt with in a mathematical fashion.

The relational malaise

It can be argued that the relational database movement squashed the possibility of data algebra for many years by a process of camouflage. It was born of good intentions. As far as we can tell from the historical record, it was a genuine attempt to introduce a data algebra – the so-called "relational algebra." However, if it ever had a sound mathematical foundation, it soon sacrificed it.

The fundamental idea underlying relational database is to organize data into tables (often called relations) of rows and columns. In general, every entity (a thing: person, product, order, invoice, etc.) has its own table and the various attributes that describe it are columns in the table. Each row in the table represents one instance of the entity (one specific person or product or whatever). In practice, not all tables represent entities, but those that do assign a unique key to each row. This is a physical implementation tactic that ensures the uniqueness of rows, and is useful for linking one entity to another (for instance linking a customer to an order). When used in this way, the unique key is described as a foreign key and is often used to join tables together.

All of this works effectively for data that fits well into tables. If relational database theory had confined itself to managing data in tables, then it is possible that a valid, but very limited, algebra could have been constructed around it. It could, perhaps, have based itself on matrix algebra (a matrix is a table or rectangular array of numbers, symbols, or expressions, arranged in rows and columns).

Unfortunately, that possibility was demolished by the introduction of SQL as a standard query language and, particularly, by the introduction of the null value, a wholly unmathematical idea. This deserves a little explanation.

Practically speaking, in a relational database, a null value can mean any of the following things:

- The value does not exist. One example might be the Social Security Number (SSN) of a U.S. resident who has not yet been issued with such a number – a newborn baby, for example.
- The value was not provided. An example could be the SSN of a U.S. resident who chooses not to provide such information.
- The value is not yet known (i.e., it could be anything). An example is the SSN of a U.S. resident who has not yet provided the database with a value for this attribute.
- The attribute is not applicable to this instance. An example is the SSN of someone who is neither a U.S. citizen nor a U.S. resident.

The first three of these situations could be kluged by assigning specific values to indicate what the context of the missing value is. And it is indeed a kluge since it changes the assigned meaning of the data item SSN. It would now mean "either an indicator denoting the status of this item or an SSN." An alternative kluge would be to assign a zero value to all three, bundling all three of these meanings together under the heading of "we do not know."

The last null possibility is far more troubling. If an entity has attributes that do not apply, then it genuinely does not have those attributes. For example, if we create a table of "living things" and include as an attribute of this table "color of beak," then many rows in the table (almost all mammals) will have null values for this attribute.

Thus, a table is the wrong structure to represent such a collection of data. If you suspect that this is a little synthetic as an example, consider instead a product table for a building materials retailer. Create such a product table, and many products will have attributes that do not apply. Paint may be sold by specific color, an attribute that is not applicable to many other products. Wood may be sold by length, thickness and width. Sand and cement may be sold by weight, and so on. There is undoubtedly an entity called product, but it cannot be represented accurately as a table. When forced under duress into a table, the table becomes populated by a great number of nulls.

If we wanted to represent this mathematically, the nulls would have to disappear, because they represent something that doesn't exist – and with their disappearance, the convenient table structure would also disappear. Instead it would become a collection of products with varying attributes.

The null is schizophrenic, possibly indicating an absent value or possibly indicating that the attribute simply does not apply. This fundamental mathematical error might not have had such a damaging impact if it were not for SQL. SQL had to deal with the nulls. In doing so, it invented a variety of rules which forced relational database further and further away from mathematical legitimacy.

These were not the only failings of the relational approach. Tables are far too limiting to be the only allowable data structures. Thus, other databases emerged (document database, XML database, RDF database, etc.) to cater to other common data structures (hierarchical data, text, graphs, etc.). However, these databases have no mathematical legitimacy either.

The consequences of bad algebra

We could make many criticisms of the relational database from a mathematical standpoint, but we think we have said enough. In the early days of the relational database movement, advocates for the technology were even emboldened to proclaim that the relational approach to data was mathematically correct, when in reality, it wasn't mathematical at all.

When data could be shoe-horned into tables, relational databases worked reasonably well. When the data didn't fit so nicely, there were problems. The idea that there is a right way to structure data (i.e., in tables) is preposterous. In truth, there are simply useful ways to structure data. Sometimes data is best structured as a graph (such as a network of relationships), sometimes as a recursive structure (such as a bill of materials), sometimes as an ordered list, and so on. Relational databases do some of these things very poorly.

Soon after relational databases became dominant, object-oriented programming languages came to the fore. These languages, which soon became quite popular, did not represent data in a relational fashion at all. This created an awkward problem when programmers chose to store data. They were obliged to translate the data as they had represented it in their program into the form the database required. This problem was called the impedance mismatch. It was so prevalent that an open source mapping capability, Hibernate, was written and widely applied to address the problem.

Thus, an IT industry emerged where programming languages and databases were incompatible in the way they modeled data. Clearly such a situation could not persist indefinitely. Nevertheless, the initial attempt to establish object databases (to complement object-oriented languages) failed commercially, and it wasn't until a second generation of object databases – this time called document databases – emerged that relational databases experienced any competition. However, these relatively new databases have no foundation in a data algebra either.

Axiomatic extended set theory

When E. F. Codd set out to formulate relational theory, he based some of his ideas on extended set theory, which was an idea formulated and described in a 1968 paper, *Description of a Set-Theoretic Data Structure* by D. L. Childs. Childs correctly identified set theory as the natural mathematical basis for representing data and proposed an extended set theory as a means to creating an algebra of data. As a consequence, it would be possible to query data using set operations such as union, intersection, Cartesian product and so on. If this were done, the use of sets and set operations would provide complete independence from physical data structures.

When Algebraix Data Corporation began work on building a data algebra, axiomatic extended set theory, as expounded by D. L. Childs, was examined to see if it was fit for this purpose. It was soon concluded that it constituted a misguided effort.

The assumption on which the genesis of data algebra moved forward was that it should be possible to derive data algebra directly from Zermelo-

Fraenkel set theory (ZFC) without need for any extension. This proved to be correct. The algebra of data described and explained in this book does indeed derive directly from ZFC. Childs' extension was unnecessary.

Nevertheless, teasing out such a universal data algebra from ZFC was not plain sailing. If it had been, an algebra of data would have been propounded decades ago, and it would quickly have become one of the foundations of software.

A final thought

Our intention in this chapter was to provide a brief review of where mathematics has been employed in software engineering at a foundational level. There are, of course, many programs and software products that implement mathematical techniques and formulae to process data (mainly numerical data) in many ways to excellent effect. Functional programming languages were developed to implement mathematical functions in a formal manner. Yet, despite the fact that data is at times even discussed in terms of sets, there was no successful attempt to implement a set theoretical approach to data – until recently.

Taking Stock

The following bullet points summarize this chapter:

- While born of the mathematical work of Alan Turing, the computer industry did not evolve mathematically. Following the computer architecture proposed by John von Neumann, it soon became the domain of electronic engineers and software engineers.
- There were some attempts to introduce and leverage mathematical ideas in software. They were as follows:
    - The introduction of functional programming languages, the most popular of which is Haskell. Functional programming languages implement a programming style that treats computation as the evaluation of mathematical functions while avoiding changing state and mutable data.
    - E. F. Codd's so-called relational algebra.
    - D. L. Childs' extended set theory.
- None of this resulted in the formulation of an algebra of data. Not even close.

Chapter 3: Cantor, Zermelo and Fraenkel

...we shall study sets, and sets of sets, and similar towers of sometimes frightening height and complexity, and nothing else.

~ Paul R. Halmos

GEORG CANTOR, a German mathematician, invented set theory single-handedly. In 1874, he unleashed it on the world by way of an article in *Crelle's Journal*. It sparked immediate controversy but very quickly had an impact, and it was soon accepted. It became very clear that set theory was fundamental to the whole discipline of mathematics. As a direct consequence, mathematicians around the world immediately set to work trying to retrofit set theory into other branches of mathematics.

As we noted in the previous chapter, Cantor's formulation of set theory was not perfect, or better put, it was not free of paradoxes. It needed to be massaged a little for the sake of rigor, so in the early twentieth century, various axiomatic systems were proposed to fix the situation.

The system suggested by Ernst Zermelo and Abraham Fraenkel was eventually adopted, and for that reason, set theory is often referred to as Zermelo-Fraenkel set theory, which may seem a little unfair given that it was Cantor's invention.

The algebra of data derives from set theory. To be precise:

The algebra of data is applied set theory!

For that reason, we have chosen here to describe and explain all the axioms of ZFC set theory on which data algebra is based. We have two motives for this.

1. We believe this will give you an opportunity to become familiar with the mathematical symbols and terminology used in set theory before we dive into the algebra of data for real.

2. These axioms are important in the derivation of data algebra, and thus it will serve you to become familiar with them. In particular, it will help if you understand what a power set is.

Mathematical symbols	
Symbol	Description
{ }	Braces indicate sets
:=	Is defined as
∈	Belongs to
∉	Does not belong to
⊂	Subset of
⊊	Subset of and not equal to
| |	Cardinality
∘	Composition
↔	Transpose
X	Cartesian product
∅	The empty set
∪	Union
∩	Intersection
'	Complement
–	Difference
▼	Cross-union
▲	Cross-intersection
▷	Superstriction
◁	Substriction
▶	Cross-superstriction
◀	Cross-substriction
⋈	Natural join

Given that we are about to introduce you to a whole series of mathematical symbols and notation, now is probably a good time to provide a complete list of every symbol used in the book in various data algebra expressions.

You may be familiar with some of these but will not be familiar with all of them. We will explain the unfamiliar ones as they arise.

Here is also an appropriate time to draw your attention to the notations we use to distinguish between various mathematical objects. We also provide a list of these in tabular form, along with font descriptions.

Notation standards		
Item	Convention	Example
element of a set	regular italic	a
couplet	regular italic with superscript	a^{b}
set	regular italic	A
set of sets	bold: font = Gauss	$\mathcal{A}$
genesis set	regular: font = Greek	Γ
power set	regular	P
sets of numbers	regular: font = X Company	$\mathbb{N}$
relation	bold italic	R
clan	regular: font = X Company	$\mathbb{C}$
function	bold italic	f
partition	regular italic: font = Greek	Π

A parade of axioms

Any game evolves subject to rules which are enforced by referees who, quite naturally, use language and terminology that comes from the rules. A casual game may have implicit rules, and it is usually refereed informally by the participants in a haphazard manner using idiomatic language. Examples of such games include hopscotch, politics and, in the authors' opinion, so-called relational algebra.

However, set theory is not a casual game: the rules – the axioms – are explicit, and the game is refereed by first order logic.

The primitive (undefined) notion of set theory is that of belonging. If an element e (whatever e is) belongs to a set E (whatever E is) we write:

$$e \in E.$$

We may also describe this notation in words by saying that "*e is an element of the set* E" or "*e is contained in E*." To assert that "*e is not an element of the set E*" we write:

$$e \notin E.$$

This may seem all hunky dory, but the very mention of E begs questions, such as "*How do we know that such a thing as a set exists,*" and "*If it does, what is it?*" Enter the axiom of existence.

Axiom of existence: *There exists a set.*

So, maybe it's E, maybe it isn't.

Axiom of extension: *Sets A and B are equal, denoted by* $A = B$, *if, and only if, they have the same elements.*

This axiom suggests the possibility of other relationships between sets aside from equality. These are the possibilities:

- If A and B are not equal, we write $A \neq B$.
- If $a \in A$ implies $a \in B$; then we say A is a subset of B (or B is a superset of A) and we write $A \subset B$. It follows that $A = B$, if, and only if, $A \subset B$ and $B \subset A$.
- If $A \subset B$ and $A \neq B$, then we write $A \subsetneq B$ and refer to A as a proper subset of B.

We illustrate these relationships in Figure 1, where the circles represent sets and the lowercase letters represent elements.

If our previously mentioned set E consists of exactly three distinct elements, say a, b and c, we use the visual artifice $\{a, b, c\}$ to represent E and we write $E = \{a, b, c\}$ (note the subtle point here that any kind of representation, such as a visual representation – which is exactly what we mean by visual artifice – is not the mathematical object, just a way of looking at it). The $\{a, b, c\}$ is the standard notation for a set. The elements are given, separated by commas, and the braces mark out the beginning and the end of the parade of elements.

We can deduce from the axiom of extension that $\{a, c, b\}$ and $\{c, b, c, a, c, a\}$ are also representations of exactly the same set E. This might not be immediately obvious, but the axiom of extension confers equality irrespective of the order of the elements. So $\{a, c, b\} = \{a, b, c\} = E$. Also, if any element is

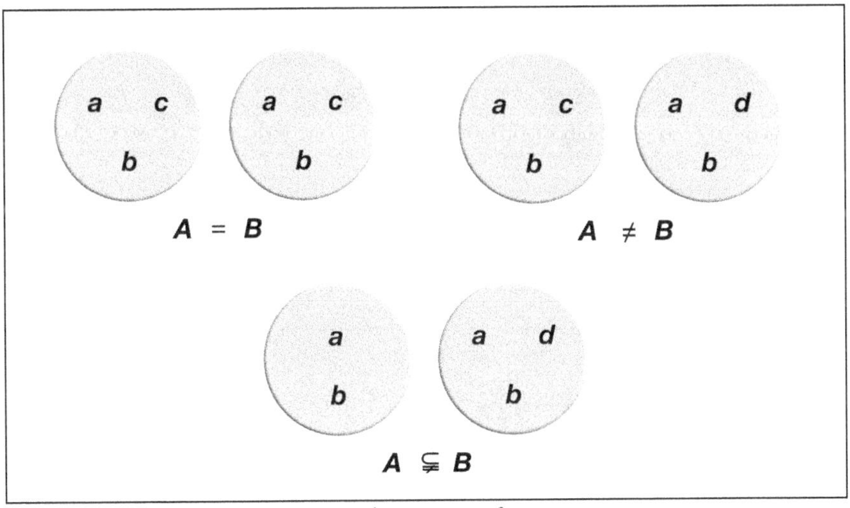

Figure 1: The axiom of extension

repeated, it still only counts once – further mentions of the element are redundant. So $\{c, b, c, a, c, a\} = \{c, b, a\} = E$. The cardinality of E (the number of elements in E) is three, and we write $|E| = 3$.

So how do we construct the subsets of a set? What are the rules?

Axiom of specification: *Every set A and every statement S(x) determines a unique subset S of A whose elements are exactly those elements x of A for which S(x) holds.*

To indicate mathematically how S is generated from A by $S(x)$ we write:

$$S = \{x \in A: S(x)\}.$$

It turns out that there are a lot of ways of mathematically specifying a statement $S(x)$. You can use atomic statements of belonging, such as $x \in E$. You can use statements of equality, such as $A = B$, and you can combine collections of such statements using the syntax of all the logical operators. The full inventory of logical operators is: $\vee$, $\wedge$, $\neg$, $\Leftrightarrow$, $\Rightarrow$, $\forall$ and $\exists$. However, we have chosen not to employ any of these in this book, preferring to use the equivalent words instead. If you do encounter any mathematical symbols you are unfamiliar with, there is a reference in Appendix A, which lists them all with descriptions. It includes every new, familiar or arcane mathematical term we employ in this book.

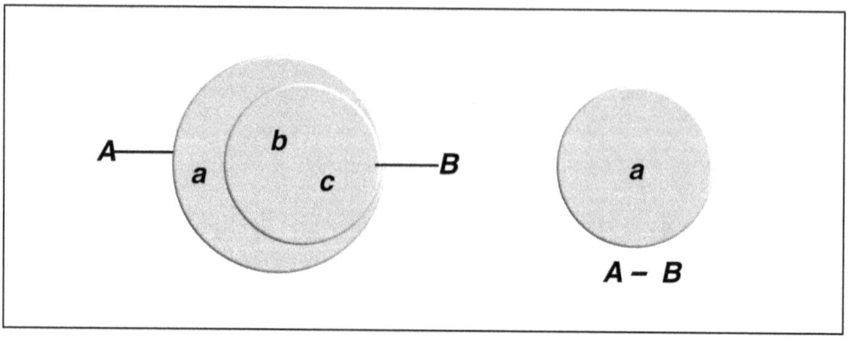

Figure 2: The difference of A and B, A – B

Consider the following example, where we use the set B to generate a set S from the set A:

$$S = A - B = \{x \in A : x \notin B\}.$$

This set S is called the difference between A and B. We have introduced "–" the difference operator, so S is the difference of the two sets, which means the set of elements that belong to A that do not belong to B.

A surprisingly useful, if somewhat slippery, set emerges as a consequence of these three axioms. Consider the set $\{x \in E : x \neq x\}$. Of course, there is no element in the whole set-theoretical universe that is not equal to itself. Nevertheless, this is not mathematical nonsense; it is an effective way of defining a set with nothing in it.

It follows from the axiom of extension that this set, which we denote by ø and refer to as the empty set, is in fact unique. Moreover, ø has the surprisingly useful property of being a subset of any set.

But hold on a moment, can a set be an element of a set? Of course it can, but we need to have a rule to state that it is so.

Axiom of pairing: *For sets A and B the set $\{A, B\}$ exists.*

This axiom allows us to play with the empty set Ø in a truly productive way.

Consider the situation where $A = B = Ø$. We can immediately, and dramatically, enlarge our inventory of sets to include:

$$\{Ø\}, \{Ø,\{Ø\}\}, \{Ø,\{Ø,\{Ø\}\}\}, \{Ø,\{Ø,\{Ø,\{Ø\}\}\}\}, \ldots$$

And this sequence of sets hints at the genesis of the set of natural numbers:

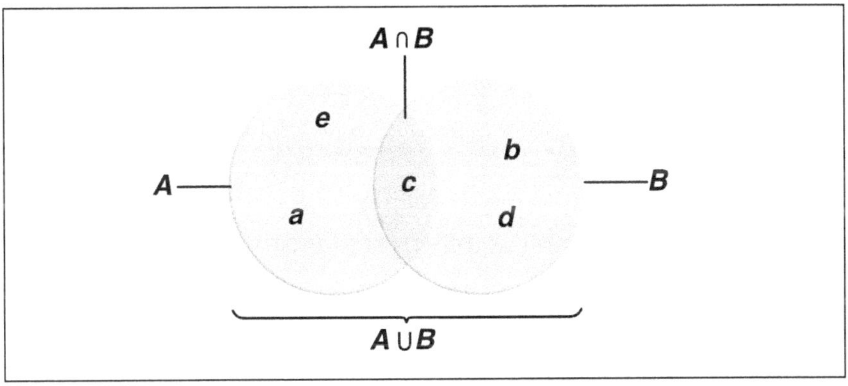

Figure 3: Union and intersection

$$\mathbb{N} := \{1, 2, 3, ...\},$$

as long as we take the ellipsis (the ... in each set) to mean "ad infinitum."

In fact, there is a mathematical method for generating the natural numbers $\mathbb{N}$ from the empty set, but we will not burden you with that. However, in case you never noticed, we slipped infinity into the mix here, so we are obliged to note that there is an axiom of infinity, which states "*there is an infinite set.*"

This leads us on to the axiom of choice, which, in simple words, states: "*one may construct a set from an infinite collection of sets by choosing an element from each set of the collection.*"

This then leads us to the axiom of substitution, which roughly states: "*anything intelligent that one can do to the elements of a set yields a set.*"

If you are starting to get the impression that we are dashing through a bad part of town hoping to get through it without mishap, then you are exactly correct. On the back of such axioms, we could digress into long, set-theoretic philosophical discussions of the implications of infinity and how it should or should not be handled. We succeed in evading that discussion with the convenient excuse that: "computers are finite, and therefore for all practical (computer) purposes, the algebra of data only needs to deal with finite sets."

So, subject to the caveat that, from time to time, it is clerically convenient to view a particular finite set as a subset of some countably infinite set – a

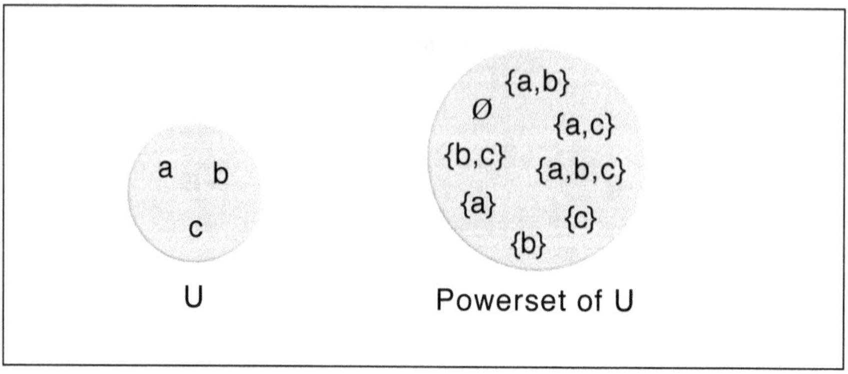

Figure 4: U and the power set of U

set whose cardinality is the same as the cardinality of $\mathbb{N}$ – we get to avoid the subtleties of infinite sets.

Let's move on to another fundamental question: "Can a set be a subset of a set other than itself?" Of course it can.

Axiom of union: *For sets A and B, there exists a set U such that* $A \subset U$ *and* $B \subset U$.

Incidentally, we generally use U to denote "the set of elements under consideration." So for example, if we are considering data about people, then we would most likely use U to denote a large population of people in a state or even in a country. It's merely a convention, of course.

The axiom of union enables us to define the union operator, ∪, by way of the axiom of extension. The axiom of extension implies that the set

$$\{ x \in U: x \in A \text{ or } x \in B \}$$

is unique. We refer to this set as the union of A and B, and we denote it by $A \cup B$. With the union of A and B in hand, we can create the subset

$$\{ x \in A \cup B: x \in A \text{ and } x \in B \}.$$

This set is referred to as the intersection of A and B, and it is denoted by $A \cap B$. And, if $A \cap B = \emptyset$, we say A and B are disjoint.

So now we have acquired two very useful binary operators for manipulating sets. The next question we can ask is:

What is the set-theoretic universe of sets that these binary operators can act within? What is the universe within which they live?

If you pursued an education that focused on science or engineering, and included a healthy dose of mathematics, it is quite possible that every set-theoretic idea we have introduced so far is not new to you. However, what we are about to describe may be new, and even if it isn't, how we are going to use this last axiom, probably is. So please take note.

Axiom of power: *For each set U, the set consisting of the subsets of U exists.*

The set of all the subsets of U is denoted by $\mathfrak{P}(U)$ and is called the power set of U. Naturally, the root set or genesis set is denoted by U to suggest "universe" because it contains all of the elements necessary to create $\mathfrak{P}(U)$.

We need to emphasize here that U is very definitely a local universe, because there is no set that contains everything. If such a set existed, it would invite the wrath of the Russell paradox (see *Shaving the barber* in Chapter 2) as it would contain an element that it does not contain!

A final note

The word "set" is a very overloaded word. It's the English word that has more definitions than any other – 464 definitions in all, according to *The Oxford English Dictionary* (if you count all variations of nouns, verbs and adjectives). One of those definitions attempts to explain the mathematical sense of the word. However, it is so misleading that we have chosen not to include it here at all. As one might suspect, other dictionaries are no better.

We tentatively suggest that if dictionary compilers feel obligated to dip their toe in the mathematical water, they adopt a definition along the lines of:

a set is a mathematical structure representing a collection of things that is rigorously defined by the ZFC set theory axioms.

Let's hope that some dictionary compiler somewhere lays their hands on this book. Admittedly, it's a faint hope.

The Power of Ascension

You may not have noticed that, so far, we haven't said much about what an algebra actually is in set-theoretic terms. Realistically, we had to cover several miles of axiomatic territory before we could do that. But now that we are armed to the teeth with the concept of the power set, we can make strides in that direction.

The word "power" in the term "power set" derives from the fact that if **U** has exactly n elements, then $\mathrm{P}(U)$ has exactly 2^n (two to the power n) elements. And, as we can also create $\mathrm{P}^2(U) := \mathrm{P}(\mathrm{P}(U))$, if we compute $2^{(2^n)}$ we will know exactly how many elements there are in the power set of the power set. And we can keep applying power set to power set as many times as we care to.

Let's take a quick look at a simple power set. If $\boldsymbol{U} = \{a, b, c\}$ then:

$$\mathrm{P}(U) = \{\emptyset, \{a\}, \{b\}, \{c\}, \{a, b\}, \{a, c\}, \{b, c\}, \{a, b, c\}\}.$$

U had 3 elements and $\mathrm{P}(U)$ has 8 (i.e., 2^3) elements. A clear advantage of the move from U to $\mathrm{P}(U)$ is the exponential increase in the number of toys in the toy box. Who could ever object to more toys, especially when the increase is from n to 2^n?

Well, to be perfectly honest, you might not be entirely happy with this when you discover you have lost some of your old favorite toys. You see:

$$\text{for every } a \in U, a \notin \mathrm{P}(U).$$

To put it bluntly, you appear to have lost all your old toys, because none of them can be found in $\mathrm{P}(U)$. But don't be too concerned; like all the Toy Story movies (so far), this is a toy story with a happy ending.

Before we reveal that, it is probably best that we define the term "ascension."

It is an important term and an important idea. We never mentioned it before because we needed you to lose all your toys in order to have a reason to explain it. If we move from operating on the set U to operating on the set $\mathrm{P}(U)$, we say that we have ascended (or lifted) from U to $\mathrm{P}(U)$. And that could be a truly positive thing if this newly minted power set provides us with extra power.

So first of all, we note that the toys we lost from $\boldsymbol{U}$ have been reincarnated in $\mathrm{P}(U)$. In other words, for every $a \in U$, $\{a\} \in \mathrm{P}(U)$. The reality is that a is not "lost" in the ascent, it is just "renamed."

Let's not forget that every element of $\mathrm{P}(U)$ is a subset of U. This turns out to be something we can leverage. Of itself, $\mathrm{P}(U)$ sprouts a natural unary operation which we refer to as complementation and define in the following way:

$$A' := U - A \quad \text{for each } A \in \mathrm{P}(U).$$

It naturally follows that $A' \cup A = U$ and $A' \cap A = \emptyset$.

But, the real power of a power set stems from the fact that it provides a richer algebraic playground than $\boldsymbol{U}$, – regardless of the nature of the elements of $\boldsymbol{U}$ – homogeneous or heterogeneous, algebraically fecund or algebraically barren. Believe it or not, by lifting from $\boldsymbol{U}$ to $\mathrm{P}(\boldsymbol{U})$ we have gotten ourselves a bona fide algebra.

It is, in fact, a set algebra with a signature:

$$[\mathrm{P}(U), \{[\cup, \emptyset], [\cap, U]\}, \{\,'\}, \{\subset\}].$$

Signatures

The purpose of a signature is to show the fundamental facts about an algebra. It's a simple idea that looks more complex than it is – what with all those brackets and braces.

Before we describe it, we need to explain the idea of an operator's identity. In simple arithmetic, the identity of the operator + is 0. Take any number, say $\boldsymbol{x}$, and add 0 to it and you get the same number. So the number 0 is said to be the identity for the operator +. Similarly, in arithmetic, the identity for the multiplication operator is 1. One of the details that an algebra's signature provides is the operators that are valid for the algebra, along with the identity (if there is one) for each operator. In general, binary operators (i.e., those which operate between two elements of a set) may have identities, but unary operators do not.

In the above signature, $[\cup, \emptyset]$ indicates that, for this algebra, the union operator is valid and its identity is the empty set $\emptyset$. So, for every set A, where $A \in \mathrm{P}(U)$, $A \cup \emptyset = A$. Similarly, for every set A, where $A \in \mathrm{P}(U)$, $A \cap \boldsymbol{U} = A$.

Given that morsel of information, the algebraic signature shown above, $[P(U), \{[\cup, \emptyset], [\cap, U]\}, \{'\}, \{\subset\}]$, can be described as follows:

> The algebra's operators shown in the signature include $\cup$ (union), $\cap$ (intersection) and $'$ (complementation). Two of these operators are binary and have identities. The third operator, $'$, is a unary operator and hence has no identity (binary operators have identities, and unary operators do not and cannot).
>
> Finally, the signature shows the relation $\subset$, which indicates the subset relationship.

Note that the signature's purpose is to highlight operations and relations that are relevant to the algebraic quest at hand. Clearly we could have mentioned the "=" relationship, but it is assumed to be part of the algebra, as it is part of every algebra. Also the signature does not inform us of all the properties of the operators that it highlights.

Nor is it necessarily the last word. As we investigate a particular algebra, we may invent a completely new operator or relation to add to the signature. As such, the signature is a living entity that may well evolve in time.

The properties of the binary operators

We now need to examine the behavior of these operators within this algebraic universe. We can start with the idea of commutativity.

We know that in arithmetic, $3 + 4 = 4 + 3$, or generally, $a + b = b + a$. This is commutativity, and we say that addition is commutative. $\cup$ and $\cap$ are clearly commutative:

$$A \cup B = B \cup A \text{ and } A \cap B = B \cap A.$$

Union and intersection are associative, just as addition is:

$$a + (b + c) = (a + b) + c.$$

Thus:

$$(A \cup B) \cup C = A \cup (B \cup C)$$

$$\text{and } (A \cap B) \cap C = A \cap (B \cap C).$$

And they have the property of mutual distributivity. With the operators $\cup$ and $\cap$:

$$(A \cup B) \cap C = (A \cap C) \cup (B \cap C)$$

$$\text{and } (A \cap B) \cup C = (A \cup C) \cap (B \cup C).$$

If you are getting the idea that the operators ∪ and ∩ are a little like the arithmetic operators + and –, that's fine, but please do not take the comparison too far, because, aside from other differences, ∪ and ∩ are also idempotent but + and – are not (and also, to tell the honest mathematical truth, and this may come as a surprise to you, – is really a unary operator[1]).

"Idempotent," incidentally is a word you will not find a great deal of use for outside mathematics. Etymologically, it comes from Latin and means "having the same power" (in Latin *idem* = the same, *potentem* = powerful). Practically, what idempotence means for ∪ and ∩ is that:

$$A \cup A = A \text{ and } A \cap A = A.$$

Of flarns, clarps and tworbles

The interaction of union, intersection and complementation is neatly captured by so-called De Morgan laws:

$$(A \cup B)' = A' \cap B' \text{ and } (A \cap B)' = A' \cup B'.$$

Stated in words, this says that:

the complement of a union is the intersection of the complements

and:

the complement of an intersection is the union of the complements.

Here we have the opportunity to introduce the nonsense words "flarn," "clarp" and "tworble." Who could resist? These words were borrowed quite a while ago from Johnny Carson's TV show by the professorial member of this book's authoring team. They are intended to have no specific meaning, aside from the fact that a flarn is not a clarp and that neither of the two is tworbles. They provide us with a useful mnemonic generalization.

Consider again: *the complement of a union is the intersection of the complements* and *the complement of an intersection is the union of the complements.* This sounds like the also true assertion from school mathematics that: *the log of a product is the sum of the logs.*

[1] *Mathematically, the expression a - b is really a shorthand for a + (-b). The unary operator "-" is applied to b, then the binary operator + is applied to both terms.*

Aren't all three of the previous italicized mantras specific versions of the more general: *the-flarn-of-the-clarp-is-the-tworble-of-the-flarns*?

Yes they are. But why are we asking?

Ascension redux

Let us revisit ascension. We need to do this because we will be using ascension several times as we gradually walk you through the levels of the tower of data algebra.

So consider our set *U*, the genesis set of elements we start out with. We can infect *U* with some algebra by taking it to be your very favorite number system from elementary school: $\mathbb{Z}$, the set of integers. The integer algebra you are familiar with is summarized by the signature

$$[\mathbb{Z}, \{[+, 0]; [\times, 1]\}; \{-\}, \{<\}].$$

In words, this signature says that the set of integers has two binary operators + and x and a unary operator – and a relationship <. We can ascend from this algebra by taking the power set of $\mathbb{Z}$. And, as we mentioned but did not prove a few pages ago, ascending from **U** to P(**U**) yields a Boolean algebra with signature:

$$[P(\mathbb{Z}), \{[\cup, \emptyset], [\cap, U]\}, \{'\}, \{-\}].$$

In view of what we already discussed, we can legitimately ask, "Have we gained set algebra at the expense of integer algebra?"

Well, yes, we have. The integers have vanished. So, $2 \notin P(\mathbb{Z})$, $3 \notin P(\mathbb{Z})$ and so on. Nevertheless, they are included in a way, because $\{2\} \in P(\mathbb{Z})$, $\{3\} \in P(\mathbb{Z})$. What is more, $\{2 + 3\} = \{5\}$ and $\{2 \times 3\} = \{6\}$. So we can choose to designate the + and x operators to work in the following way:

$$\{2\} + \{3\} = \{5\} \text{ and } \{2\} \times \{3\} = \{6\}.$$

Be clear about what we have done here. The operators + and x in the integer algebra we started with are not the same as the operators + and x in the Boolean algebra of P($\mathbb{Z}$). We have "overloaded" these two symbols. But it doesn't cause any problems, because $\{\{i\} \in P(\mathbb{Z}): i \in \mathbb{Z}\}$ perfectly replicates the algebra of $\mathbb{Z}$ in P($\mathbb{Z}$).

It may be clearer and perhaps even instructive to translate this into words: *the overloaded sum (product) of two singleton sets of integers is the singleton set of the integer sum (product) of the integers.*

This generalizes in a convenient way to the elements of $P(\mathbb{Z})$, which are not necessarily singleton sets: for $A, B \in P(\mathbb{Z})$

$$A + B := \{a + b \in \mathbb{Z}: a \in A \text{ and } b \in B\}.$$

So: *the sum of sets is the set of sums* (a version of our mantra, *the-flarn-of-the-clarp-is-the-clarp-of-the-flarns*), and

$$A \times B := \{a \times b \in \mathbb{Z}: a \in A \text{ and } b \in B\}.$$

And: *the product of sets is the set of products* (again, *the-flarn-of-the-clarp-is-the-clarp-of-the-flarns*).

Let's take a look at some examples of this:

$$\begin{aligned}\{1, 3\} + \{4, -1, 0\} &= \{1 + 4, 1+ (-1), 1 + 0, 3 + 4, 3+(-1), 3 + 0\} \\ &= \{5, 0, 1, 7, 2, 3\}\end{aligned}$$

and:

$$\begin{aligned}\{1, 3\} \times \{4, -1, 0\} &= \{1 \times 4, 1 \times (-1), 1 \times 0, 3 \times 4, 3 \times (-1), 3 \times 0\} \\ &= \{4, -1, 0, 12, -3, 0\}.\end{aligned}$$

It follows that $P(\mathbb{Z})$ is an "arithmetic" algebra in its own right, constructed by lifting the operations of $\mathbb{Z}$ to $P(\mathbb{Z})$, with signature

$$[P(\mathbb{Z}), \{[+, \{0\}], [\times, \{1\}]\}, \{-\}].$$

Right now it may look to you as though we've pulled off a dramatic ascension and lifted the whole of arithmetic algebra into $P(\mathbb{Z})$. But actually, we have not.

We have lifted much of the structure of integer arithmetic but not all of it. On the positive side, for $A, B, C \in P(\mathbb{Z})$ the lifted versions of + and $\times$ are commutative, associative and $\times$ distributes over +. Of course, the lift of + does not distribute over $\times$ in $P(\mathbb{Z})$, but it didn't do that in $\mathbb{Z}$ either. Indeed, it looks as though we've tamed the whole beast; that is, until the beast suddenly goes feral.

Consider: $-\{1, 2\} + \{1, 2\}$. Surely that's equal to $\{0\}$. Well, not exactly...

$$-\{1, 2\} + \{1, 2\} = \{-1, -2\} + \{1, 2\}$$
$$=\{-1+1, -1+2, -2+1, -2+2\}$$
$$=\{0, 1, -1, 0\}$$
$$=\{0, 1, -1\}.$$

Quite clearly:

$$-\{1, 2\} + \{1, 2\} \neq \{0\}.$$

This means a whole cancellation law goes missing when you lift to P($\mathbb{Z}$). In other words, if $-A + A \neq \{0\}$, then $A + B = A + C$ does not imply that $B = C$. You cannot cancel the A from each side of the equation.

Nevertheless, the benefits of the lifting process far outweigh such anomalies. In this instance, if we look hard enough, we will discover the complete inventory of modular arithmetics in P($\mathbb{Z}$)!

The modular arithmetic of the power set P($\mathbb{Z}$)

You learned modular arithmetic, at least in a primitive way, when you were a child and learned how to read the time from a clock face. If someone asked you at 11 o'clock what time it would be in three hours, you swiftly answered "2 o'clock," adding 3 to 11 and subtracting 12. You performed modulo 12 addition.

So let's consider the simplest modular arithmetic of all: arithmetic modulo 2. And let's denote all the even integers by E and all the odd integers by O. We can confidently state that $E, O \in$ P($\mathbb{Z}$).

Why? Well, P($\mathbb{Z}$) is a power set that contains all possible subsets of $\mathbb{Z}$. Both the sets E *and* O must be in there somewhere.

Moreover, $E \cap O = \emptyset$ and $E \cup O = \mathbb{Z}$. So E *and* O partition the set $\mathbb{Z}$. Now let's focus on the subset $\{E, O\}$ of P($\mathbb{Z}$) and start employing our operators + and x. Add an even number to an even number, you get an even number. Add odd to odd and you get even, and so on. The full set of possibilities for + and x is shown in the two tables.

+	E	O
E	E	O
O	O	E

x	E	O
E	E	E
O	E	O

So take a look at these two tables and then replace E with 0, which suggests that the common remainder when any even integer is divided by two is 0, and replace the O with 1, suggesting that the common remainder of any odd integer divided by two is the integer 1.

+	0	1
0	0	1
1	1	0

x	0	1
0	0	0
1	0	1

This reveals integer arithmetic modulo 2 hiding shamefully in P($\mathbb{Z}$), with a signature that looks like this:

$$[\{0, 1\}, \{[+, 0], [x, 1]\}]$$

In fact, the integer arithmetic modulo any-other-positive-integer-of-your-choice, including clock arithmetic, is hiding in P($\mathbb{Z}$) somewhere.

The importance of partitions

Each of these modular algebras depends on a particular partition of Z. This is not an anomaly; partitions play a singular and prominent role in mathematics. They have a role to play in the conventional algebra you met at school and grew to love or hate, and they play a role in more general and more abstract algebras. And as you must realize, otherwise why would we even mention them, they play a role in data algebras.

So it's convenient for all of us that a partition is not a complex thing. Still, this is a game with rules, so a precise definition is in order:

> A **partition of a set** U is a set, say, Π, of non-empty subsets of U such that the intersection of distinct elements of Π is empty and the union of all of the elements of Π is U.

And by the way, an element of a partition is referred to as a component of the partition.

For example, if U is {1, 2, 3, 4, 5}, then {{1, 2, 3}, {3, 4, 5}} is not a partition of U but {{1, 2}, {3, 4, 5}} is, with components {1, 2} and {3, 4, 5}. Notice also that a partition of U is a subset of P(U), and, by the way, that means it must be an element of $P^2(U)$.

Taking Stock

In this chapter we covered a good deal of mathematical territory – all of it foundational to data algebras. The following bullet points summarize our path through that territory.

- The algebras of data are applied ZFC set theory (incidentally, the ZFC stands for Zermolo-Fraenkel Set Theory with the axiom of Choice).
- ZFC set theory is established by nine largely self-evident axioms. The axioms of existence, extension, specification, pairing, infinity, choice, substitution, union and power. The axiom of power – that we can form a power set consisting of all possible subsets of a local set universe – is of specific importance to data algebra.
- In set theory, the fundamental operators are union ($A \cup B$), and intersection ($A \cap B$), the unary operator of complementation ', where $A' = U - A$. There is also a fundamental subset relation $\subset$.
- Binary operators may have specific properties of commutativity, associativity, distributivity (between two binary operators) and idempotence. $\cup$ and $\cap$ are commutative, associative, mutually distributive and idempotent.
- By taking the power set P(U) of a universe, U, we can construct a Boolean algebra. Algebras can be represented by signatures, and the signature of this specific Boolean algebra is:

  $$[\mathrm{P}(U), \{[\cup, \emptyset], [\cap, U]\}, \{\,'\,\}, \{\subset\}].$$

  Thus, we can lift U to P(U).
- In lifting from an algebra to its power set we can always lift the operations of the algebra to the power set, but the algebraic properties of the lifted operations may not necessarily follow in kind. This is the case in lifting the unary operator "–" from $\mathbb{Z}$ to P($\mathbb{Z}$).
- Nevertheless, in lifting $\mathbb{Z}$ to P($\mathbb{Z}$) we discover that the whole of modular arithmetic resides within P($\mathbb{Z}$) with each specific modular arithmetic depending on a particular partition of Z.
- A partition of a set U is a set, say, Π, of non empty subsets of U such that the intersection of distinct elements of Π is empty and the union of all of the elements of Π is U.

- An element of a partition is referred to as a component of the partition.
- A partition of U is a subset of $P(U)$, which means it must also be an element of $P^2(U)$.

Chapter 4: Descartes' Legacy

Obvious is the most dangerous word in mathematics.

~ E. T. Bell

IN ACCORD with the academic fashion of the day, René Descartes' name was latinized to Renatus Cartesius. That, believe it or not, is why the adjective from his surname is "Cartesian" and not "Descartesian." No doubt you encountered this adjective in school mathematics, as geometry is usually introduced in terms of algebraic functions using so-called Cartesian coordinates.

This chapter commences with that, but what it really focuses on is Cartesian products, which you may not be so familiar with. However, we can commence by stumbling into some graphs that ought to inhabit your mathematical comfort zone.

"Ordered pairs"

Take a look at Figure 5. Perhaps you think: "OK, it's got a y-axis and an x-axis, and it's got some scattered Cartesian coordinates marking out some points on the two-dimensional surface referred to grandiosely, by mathematics geeks, as the Euclidean plane. And that is just a mathematics geek's playground for drawing graphs relating y to x in some obscure way and marking the resulting curve out with 'ordered pairs' plucked from ℝ, the set of real numbers."

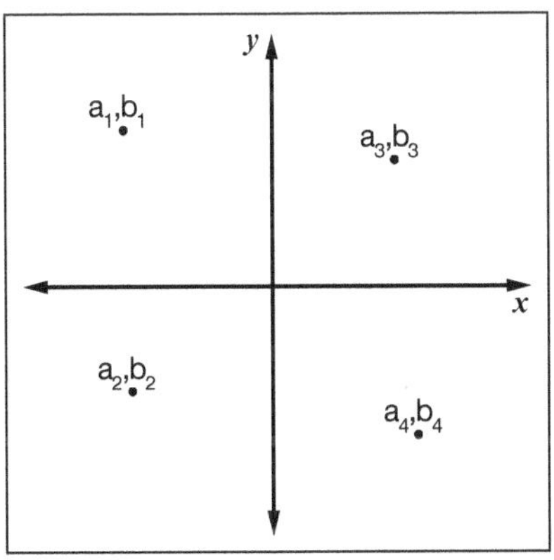

Figure 5: Cartesian coordinates

Maybe you even recall the school lesson where such a graph over the Euclidian

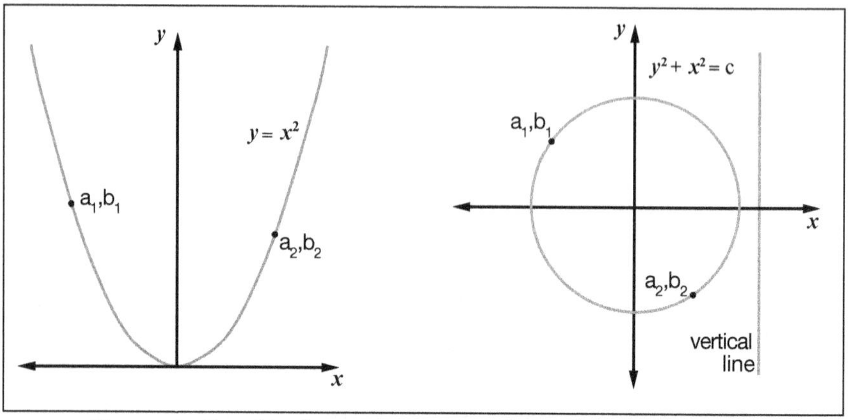

Figure 6: Graphical representations of $y = x^2$ and $y^2 + x^2 = c$.

plane was first unveiled before your eyes, most probably in an effort to illustrate simple equations such as $y=2x$ and $y=x^2$. Perhaps you even recall the "vertical line test" from your school days, which is illustrated in Figure 6. The test is this: *if each vertical line that can be drawn in the plane intersects a curve in at most one point (an "ordered pair") the curve represents a relation between* **y** *and* **x**, *which is referred to as a function.* If that is not so then, as in the diagram above, you have drawn a relation that is not a function.

In some sense, functions were the gold standard of your secondary mathematics courses, running like a glittering thread through much of what you learned. However, this is not so when it comes to the algebra of data. In fact, hopefully, you will gradually realize that relations rather than functions are the gold standard of data algebra. Relations are gloriously flexible, and their flexibility steps forward to address data algebra's need to handle ambiguity and generality.

Now is as good a time as any to get a dark and ugly secret out into the open. If you don't already know it, then all we can say is: we hope you're ready for it...

> *Most of the relations one encounters in applying mathematics outside of the classroom do not lend themselves to such pretty closed-form expressions as $y = x^2$ and $y^2 + x^2 = c$.*

This is true in spades for data algebra. For most of the time, we will have to be satisfied with exhibiting relations as a set (of very large cardinality in practice) of "ordered pairs." You see, our data will consist of entities which aren't typically thought of as mathematical objects, even though they can indeed be represented as such objects.

By the way, have you asked yourself yet why we persist in writing "ordered pair" in quotes? If you have, good. If you haven't, do it now, and if you are up for it, try to guess, before you start reading about Kuratowski's definition of the "ordered pair." But for a few pages before that, we need to explain Cartesian products.

Cartesian products

The Cartesian product of two sets, say A and B, is a set-theoretic construction that depends on our ability to make the ascent from $A \cup B$ to $P^2(A \cup B)$, and then carefully, very carefully, as if performing brain surgery, selecting a certain subset of $P^2(A \cup B)$, which we denote by $A \times B$ and refer to as the Cartesian product of A and B. Perhaps you never quite understood that (and if you did, well done!), but don't worry, it will gradually become clear. For now, be content with the important knowledge that Cartesian products spawn the relations that make algebras tick.

Let A and B be sets. The Cartesian product of A and B is defined by:

$$A \times B := \{\{\{a\}, \{a, b\}\} \in P^2(A \cup B) : a \in A \text{ and } b \in B\}.$$

We refer to an element, $\{\{a\}, \{a, b\}\}$, of a Cartesian product as a couplet. In the short term, we will use the visual artifice (a, b), with which you are familiar, as a stand-in for the notationally more cluttered $\{\{a\},\{a, b\}\}$. More on this later, when you will be hearing a lot about couplets.

Importantly, for our purposes, we note that the Cartesian product $\times$ distributes over the union operator $\cup$ as follows:

$$(A \cup B) \times (A \cup B) = (A \times A) \cup (A \times B) \cup (B \times A) \cup (B \times B).$$

If we leave that floating around in the middle of the page, you probably won't see what's happening, so we will employ a visual artifice to assist.

A good way to picture a Cartesian product is as a table, with each couplet occupying a cell in the table. So the Cartesian product of $A = \{2, 4, 6\}$ with $B = \{1, 3, 5\}$ can be illustrated as follows:

$A \times B$		B		
		1	3	5
A	2	(2,1)	(2,3)	(2,5)
	4	(4,1)	(4,3)	(4,5)
	6	(6,1)	(6,3)	(6,5)

$A \times B = \{(2,1), (2,3), (2,5), (4,1), (4,3), (4,5), (6,1), (6,3), (6,5)\}$.

Incidentally, there is nothing to stop us from creating the Cartesian product of sets with different cardinalities, say $A = \{2, 4\}$, $B = \{1, 2, 3, 4\}$, as illustrated in the table below:

$A \times B$		B			
		1	2	3	4
A	2	(2,1)	(2,2)	(2,3)	(2,4)
	4	(4,1)	(4,2)	(4,3)	(4,4)

$A \times B = \{(2,1), (2,2), (2,3), (2,4), (4,1), (4,2), (4,3), (4,4)\}$.

So with $A = \{2, 4\}$ and $B = \{1, 4\}$, and let us now demonstrate that:

$$(A \cup B) \times (A \cup B) = (A \times A) \cup (A \times B) \cup (B \times A) \cup (B \times B).$$

$A \times A$		A	
		2	4
A	2	(2,2)	(2,4)
	4	(4,2)	(4,4)

$A \times A = \{(2,2), (2,4), (4,2), (4,4)\}$.

$A \times B$		B	
		1	4
A	2	(2,1)	(2,4)
	4	(4,1)	(4,4)

$A \times B = \{(2,1), (2,4), (4,1), (4,4)\}$.

Note, as we shall now illustrate, $A \times B$ is definitely not the same as $B \times A$ (i.e., $A \times B \neq B \times A$):

$B \times A$		A	
		2	4
B	1	(1,2)	(1,4)
	4	(4,2)	(4,4)

$B \times A = \{(1,2), (1,4), (4,2), (4,4)\}$.

See – definitely not the same as $A \times B$. So, on to $B \times B$:

$B \times B$		B	
		1	4
B	1	(1,1)	(1,4)
	4	(4,1)	(4,4)

$B \times B = \{(1,1), (1,4), (4,1), (4,4)\}$.

Stringing it all together,:

$$(A \times A) \cup (A \times B) \cup (B \times A) \cup (B \times B) = \{(2,2), (2,4), (4,2), (4,4), (2,1), (2,4), (4,1), (4,4), (1,2), (1,4), (4,2), (4,4), (1,1), (1,4), (4,1), (4,4)\}$$

$$= \{(1,1), (1,2), (1,4), (2,1), (2,2), (2,4), (4,1), (4,2), (4,4)\}$$

Now consider $(A \cup B) \times (A \cup B)$. Let $C = A \cup B$.

Then $C = \{2,4\} \cup \{1,4\} = \{1, 2, 4\}$. So $C \times C$ is given by:

$C \times C$		C		
		1	2	4
C	1	(1,1)	(1,2)	(1,4)
	2	(2,1)	(2,2)	(2,4)
	4	(4,1)	(4,2)	(4,4)

So, $(A \cup B) \times (A \cup B) = (A \times A) \cup (A \times B) \cup (B \times A) \cup (B \times B)$ is true in this example, at the very least. And, of course, it is true in general.

We can thus assume that any set of couplets is a subset of a Cartesian product of the form $U \times U$, for an appropriately chosen U. Having noted that $A \times B \neq B \times A$, we need to state the relationship more formally, by writing:

$$A \times B = B \times A \text{ if, and only if, } A = \emptyset \text{ or } B = \emptyset \text{ or } A = B.$$

and also:

$$A \times (B \times C) = (A \times B) \times C \text{ if, and only if, } A = \emptyset \text{ or } B = \emptyset \text{ or } C = \emptyset.$$

If you remember the terms "commutative" and "associative," you can now confidently declare that the Cartesian product is neither commutative nor associative.

In reality, the expression $A \times B \times C$ is as dubious as the arithmetic expression $6 - 3 - 7$ because there is no indication of which operator to apply first. Indeed, $A \times B \times C$ is notational nonsense unless $A = \emptyset$ or $B = \emptyset$ or $C = \emptyset$.

Heavy lifting

In Chapter 3, the arithmetic ascent from $\mathbb{Z}$ to $P(\mathbb{Z})$ served as our introduction to lifting, happily lending itself to basic notation:

"For $A, B \in P(\mathbb{Z})$, $A + B := \{a + b \in \mathbb{Z}: a \in A \text{ and } b \in B\}$",

and smooth talk:

"The sum of sets of integers is the set of the sums."

Recall, as we once told you, that the sequence:

$$\{\emptyset\}, \{\emptyset,\{\emptyset\}\}, \{\emptyset,\{\emptyset,\{\emptyset\}\}\}, \{\emptyset,\{\emptyset,\{\emptyset,\{\emptyset\}\}\}\},$$

is the genesis of the set of natural numbers:

$$\mathbb{N} := \{1, 2, 3,...\}.$$

True, we didn't prove that mathematically when we declared it so in Chapter 3, and neither are we going to prove it now. Nevertheless, please take our word for it.

Your parents (or maybe just your mother) knew about $\mathbb{N}$, because early on they (or she) turned $\mathbb{N}$ into the "counting" algebra $[\mathbb{N}; \{[+]\}]$ for you. No frills, such as zero or negation or even philosophical contemplation of the meaning of infinity; just counting, adding, commuting and associating with reckless abandon.

But, had your mom revealed to you the rudiments of set theory, then maybe, as a precocious mathematical child, together you would have explored the Cartesian product of this "counting algebra." You would surely have realized the sense in doing so, because your mom would doubtless have advised you that if your algebra seems somewhat limited, then you should ascend to a Cartesian product, like all precocious kids do.

So if you've had a mathematically deprived childhood, take the time now to ascend to the Cartesian product:

$$\mathbb{N} \times \mathbb{N} = \{(a, b) \in P2(\mathbb{N}): a, b \in \mathbb{N}\}$$

and amuse yourself by meandering around. Really, take a look:

Nx N	N							
N	...	...	...	...	...	...	...	...
	(1,7)	(2,7)	(3,7)	(4,7)	(5,7)	(6,7)	(7,7)	...
	(1,6)	(2,6)	(3,6)	(4,6)	(5,6)	(6,6)	(7,6)	...
	(1,5)	(2,5)	(3,5)	(4,5)	(5,5)	(6,5)	(7,5)	...
	(1,4)	(2,4)	(3,4)	(4,4)	(5,4)	(6,4)	(7,4)	...
	(1,3)	(2,3)	(3,3)	(4,3)	(5,3)	(6,3)	(7,3)	...
	(1,2)	(2,2)	(3,2)	(4,2)	(5,2)	(6,2)	(7,2)	...
	(1,1)	(2,1)	(3,1)	(4,1)	(5,1)	(6,1)	(7,1)	...

Make a careful note of what we are about to do. Because we are about to demonstrate an ascension up the algebraic tower from $\mathbb{N}$ to $\mathbb{Z}$, using nothing more than the above Cartesian product – absolutely no smoke and no mirrors.

If you really take time studying this Cartesian product you might discover:

- Since we can add natural numbers, we can certainly lift + to $\mathbb{N} \times \mathbb{N}$, by defining the sum of couplets to be the couplet of sums: $(a, b) + (c, d) := (a + c, b + d)$: For example (2, 3) + (4, 1) = (6, 4). So that gives us the algebra $[\mathbb{N} \times \mathbb{N}; \{+\}]$. And if you check, you will find that this lifted addition is pleasingly commutative and associative.
- Maybe the "diagonals" of our $\mathbb{N} \times \mathbb{N}$ array have caught your attention. If you follow the "main" diagonal traveling from bottom left to top

right, it is populated by couplets of the form (x, x). The "higher" diagonal above, starting with (1, 2) has couplets $(x, x+1)$. In general, the higher k-th diagonal has couplets $(x, x+\mathrm{k})$ and the lower k-th diagonal has couplets that are the transpose of $(x, x+\mathrm{k})$. i.e., $(x+\mathrm{k}, x)$, where $\boldsymbol{x}$ ranges over $\mathbb{N}$.

- What does this portend? Are the higher diagonals positive in some sense and the lower diagonals negative in some sense? What does this say about the main diagonal? Could that be zero? And is the diagonal in fact a mirror? Corresponding higher and lower diagonals are indeed reflections of each other through the main diagonal.
- Does it look as though we have found the set of integers $\mathbb{Z}$ hiding playfully in the diagonals of couplets of $\mathbb{N}$ **x** $\mathbb{N}$?

Wait a minute. We are getting ahead of ourselves by designating these diagonals to be "entities" in their own right. Remember, $\mathbb{N}$ **x** $\mathbb{N}$ is a set of couplets, not a set of diagonals. The diagonals are merely a visual feature in a visual artifice and have no standing of themselves. Nevertheless, each diagonal is undeniably a set of couplets. And that means that each diagonal can be properly viewed as an element of the power set P($\mathbb{N}$ **x** $\mathbb{N}$).

So watch closely, while we perform some genuine mathematical manipulations, and try not to get the impression that you're playing Three Card Monte with a street-wise card shark.

To maintain set-theoretic integrity, we must ascend to P($\mathbb{N}$ **x** $\mathbb{N}$) and identify the elements of the subset, let us call it $\mathbb{D}$, of P($\mathbb{N}$ **x** $\mathbb{N}$) whose elements are our diagonals, by numerals in bold font.

For example, 4 denotes:

$$\{(\mathrm{x}, \mathrm{x}+4)\} \in \mathbb{N} \mathbf{x} \mathbb{N}: \mathrm{x} \in \mathbb{N}\}.$$

It follows that the transpose of 4, let's represent it as $\overleftrightarrow{4}$, denotes:

$$\{(x+4, x)\} \in \mathbb{N} \mathbf{x} \mathbb{N}: x \in \mathbb{N}\}.$$

We "invent" the symbol 0 to denote:

$$\{(x, x)\} \in \mathbb{N} \mathbf{x} \mathbb{N}: x \in \mathbb{N}\},$$

because this diagonal, and only this one, is not characterized by any element of $\mathbb{N}$.

We now lift "+" yet again, and restrict our attention to $\mathbb{D}$. This yields:

$$\begin{aligned}\mathbf{4} + \overleftrightarrow{\mathbf{4}} &= \{(x, x{+}4) + (y{+}4, y) \in \mathbb{N} \mathbf{x} \mathbb{N} : x,\ y \in \mathbb{N}\}\\ &= \{(\mathrm{x{+}y{+}4,\ x{+}4{+}y}) \in \mathbb{N} \mathbf{x} \mathbb{N}\text{: x, y} \in \mathbb{N}\}\\ &\subset \{(z, z) \in \mathbb{N} \mathbf{x} \mathbb{N}\text{: } z \in \mathbb{N}\}.\end{aligned}$$

In other words, each possible sum of elements from components **4** and $\overleftrightarrow{\mathbf{4}}$ of our partition $\mathbb{D}$ of $\mathbb{N} \times \mathbb{N}$ is the component **4** of $\mathbb{D}$. So, by unanimous democratic vote, $\mathbf{4} + \overleftrightarrow{\mathbf{4}} := 0$.

Similarly:

$$\begin{aligned}\mathbf{0} + \mathbf{4} &= \{(x, x) + (y, y{+}4) \in \mathbb{N} \mathbf{x} \mathbb{N}\text{: } x,\ y \in \mathbb{N}\}\\ &= \{(x{+}y, x{+}y{+}4) \in \mathbb{N} \mathbf{x} \mathbb{N}\text{: } x,\ y \in \mathbb{N}\}\\ &\subset \{(z, z{+}4) \in \mathbb{N} \mathbf{x} \mathbb{N}\text{: } x, y \in \mathbb{N}\}.\end{aligned}$$

Thus, each possible sum of elements from components 0 and 4 of our trusty partition of $\mathbb{N} \times \mathbb{N}$ is the component **4** of $\mathbb{D}$. By overwhelming vote: $\mathbf{0} + \mathbf{4} := \mathbf{4}$.

And, similarly:

$$\begin{aligned}\mathbf{4} + \overleftrightarrow{\mathbf{5}} &= \{(x, x{+}4) + (y{+}5, y) \in \mathbb{N} \times \mathbb{N}\text{: } x,\ y \in \mathbb{N}\}\\ &= \{(x{+}y{+}5, x{+}4{+}y) \in \mathbb{N} \times \mathbb{N}\text{: } x,\ y \in \mathbb{N}\}\\ &= \{((x{+}y{+}4) + 1), x{+}4{+}y) \in \mathbb{N} \times \mathbb{N}\text{: } x,\ y \in \mathbb{N}\}\\ &\subset \{(z{+}1, z) \in \mathbb{N} \mathbf{x} \mathbb{N}\text{: } z \in \mathbb{N}\} = \overleftrightarrow{1}.\end{aligned}$$

And so, each possible sum of elements from component 4 and component $\overleftrightarrow{5}$ respectively, of our partition of $\mathbb{N} \times \mathbb{N}$ is an element of the component $\overleftrightarrow{1}$ of $\mathbb{D}$. The votes are in, and the vote count indicates a landslide for:

$$\mathbf{4} + \overleftrightarrow{\mathbf{5}} = \overleftrightarrow{\mathbf{1}}.$$

Lo and behold, to your complete mathematical astonishment, it turns out that $\mathbb{D}$, an infinite set of infinite sets of couplets, together with lifted addition, lifted multiplication and lifted transposition is the authentic man-made integer algebra, which signs its checks in the following way:

$$[\mathbb{D}, \{[+, 0], [\mathbf{x}, 1], \{\leftrightarrow\}].$$

Now if you are going to quibble about the fact that we only demonstrated addition and not multiplication, just take our word that it's not sleight of hand that's in play here. The simple fact is that proving out multiplication in a similar manner to the above is very messy and thus, in time-honored fashion, we leave it as an exercise for the reader.

By the way, it was $\mathbb{D}$'s doppelganger, whose autograph is:

$$[\mathbb{Z}, \{[+, 0], [\times, 1], \{-\}\}],$$

that you encountered long ago in elementary school.

If you gaze at this famous signature and wonder to yourself how it could be that the "$\leftrightarrow$" transpose operator of $\mathbb{D}$ is the same as the "$-$" minus operator of $\mathbb{Z}$, we can only console you with the words: the truth is that they do exactly the same thing.

And if the ingrained learned-by-rote mantra "the negative of a negative is a positive" suddenly pops up in your consciousness at this point, you can append "and the transpose of a transpose is what you started with in the first place," knowing full well that they indicate exactly the same thing.

The fact is that your journey through school mathematics was a voyage into number systems of ever increasing computational richness and cardinality: from the naturals, to the integers, to the rationals, to the reals and then on to the complex numbers.

Most likely the impressive-looking relationships:

$$\mathbb{N} \subset \mathbb{Z} \subset \mathbb{Q} \subset \mathbb{R} \subset \mathbb{C}$$

appeared on the blackboard or the whiteboard at some point, as a summary of this progression. Sorry we have to point it out, but this impressive chain of relationships is not exactly kosher, or perhaps better put, it is notationally misleading.

For example, we know that the naturals are not a subset of the integers. As we have just demonstrated, it would be far more accurate to proclaim that there is a proper subset of the algebraically richer integers, consisting of those upper diagonals we constructed, that is algebraically equivalent to the natural numbers.

The upshot of this is not that we should curse our school mathematics teachers, who are anyway obliged to take one or two mathematical liberties in order to introduce us to foundational mathematical ideas. No. The real lesson is that our investment in a Cartesian product has yielded considerable algebraic profit.

And so it proceeds by numeric ascent. The rationals $\mathbb{Q}$ are derivative of $\mathbb{Z}$ and contain a clone of the integers. The reals $\mathbb{R}$ are derivative of $\mathbb{Q}$ and contain a clone of the rationals. The complex numbers $\mathbb{C}$ are derivative of $\mathbb{R}$ and contain a clone of the reals.

Here we can point at a whole numeric tower of algebraic ascension, supported by a very helpful scaffolding of Cartesian products. This series of lifts from one algebra to another was essential to the development of modern mathematics.

The same turns out to be true in the development of data algebras. There is a tower, there are Cartesian products and there is an elevator to help us ascend.

The Kuratowski Diversion

We would rather not involve you in the debates and deliberations of pure mathematicians, but we feel obliged to do so in the case of the "ordered pair." That terminology and the traditional visual artifice married to it, once out of the bag, fairly engenders unwelcome ambiguity.

Our story begins with Kazimierz Kuratowski and his set construct {{a}, {a,b}}. Kuratowski offered this construct as the definition of an "ordered pair" in 1921, and it was generally accepted. So far, so good, in the sense that it simultaneously binds and distinguishes two distinct elements in one entity. Moreover, if the elements are not distinct (i.e., $a = b$), the issues of binding and distinguishing are moot, which is captured by:

$$\{\{a\}, \{a, a\}\} = \{\{a\}, \{a, a\}\} = \{\{a\}, \{a\}\} = \{\{a\}\}.$$

If this definition is not peculiar to a single element or a pair of distinct elements, then it begs for a generalization that maintains the set-theoretic integrity of the original while extending its reach up to as many as three distinct elements. The seemingly natural choice for an "ordered triple:"

$$(a, b, c) := \{\{a\}, \{a, b\}, \{a, b, c\}\},$$

appears promising: (a, b) and (a, b, c) are fellow members of $P^2(U)$, and this fellowship is shared by so-called "ordered n-tuples" that are defined in this manner. That's fine, at first blush, but it gives birth to anomalies. For example:

$$(a, a, a) = \{\{a\}, \{a, a\}, \{a, a, a\}\} = \{\{a\}\} = (a, a)$$

and:

$$(a, a, c) = \{\{a\}, \{a, a\}\ \{a, a, c\}\} = \{\{a\}, \{a, c\}\} = (a, c).$$

If you don't immediately see something amiss with these equations, then think of the Cartesian coordinates that we know and love. An "ordered triple" implies three axes, or dimensions if you prefer, and will not settle for two. Now let us pause while you work yourself into a frenzy at the mathematical atrocity, implicit in these equations......

It is possible to deal with this and other problems by bending the rules – in effect by adding new axioms to the game of ZFC. But this is a price that we are not willing to pay for several reasons, one of which is that it would hinder our effort to treat with data in its full generality.

So what to do?

We shall ignore the begging, and never again in this text or in our description of data algebra use the term "ordered pair" or the visual artifice (a, b) in reference to the Kuratowski construct $\{\{a\}, \{a, b\}\}$. A couplet is a couplet is a couplet: a certain element of the second power set of some set of your choosing. By making no more of a couplet than that, we are able to make much more of couplets than that.

A visual artifice for the couplet

You probably realize from the above that the couplet plays a significant role in data algebra, otherwise why would we have troubled you with discussion of its definition? In addition to substituting the term "couplet" for the term "ordered pair," we shall now introduce some new mathematical notation.

The reason is this. In a chapter or two, we shall be obliged to discuss sets of couplets, sets of sets of couplets and so on. Imagine how that's going to look if we retain the braces of $\{\{a\}, \{a, b\}\}$ – a forest of braces all brawling with each other to define where we are and what mathematical object we are dealing with in our "towers of sometimes frightening height and complexity."

While we could have proceeded instead with the notation (a, b), we also judged this to be a pair of parentheses too far. In the end, we adopted the exponential form, deciding to use a^b in place of $\{\{a\}, \{a, b\}\}$. Our rationale was:

- a^b is less cluttered and more humane than the tortuous $\{\{a\}, \{a, b\}\}$.
- a^b visually captures the coalescence of a and b, while at the same time distinguishing them.
- a^b suggests a symmetric clone, b^a, which turns out to be important and which we will later refer to as the transpose of a^b.
- a^b and b^c almost beg to be composed; that is $a^b \circ b^c = a^c$, which indeed turns out to be a fertile binary operation in data algebras.

Now, at the very end of this chapter, it is time to introduce an important mathematical question:

What does it mean for a^b and x^y to represent the same couplet?

The facile answer, based on nothing more than the fact that a^b and x^y are just visual artifices, is that $a = x$ and $b = y$. Such a response is not only mathematically insufficient; to dyed-in-the-wool mathematicians, it is mathematically abhorrent. The point is that visual artifices do nothing more than represent mathematical objects, in this case, couplets, and if we are going to prove that $a^b = \boldsymbol{x}^{\boldsymbol{y}}$ implies that $a = \boldsymbol{x}$ and $b = \boldsymbol{y}$, we can only do so by appealing to actual mathematical entities:

$$\text{“}\{\{a\},\{a,b\}\} \text{ and } \{\{x\},\{x,y\}\}\text{.”}$$

To put it briefly, mathematics always dictates to visual artifices, whether they happen to be diagrams (such as Venn diagrams) or textual representations, and the vice is not versa.

Your mission, should you wish to accept it, is to prove that:

$$a^b = x^y \text{ if, and only if, } a = x,\ b = y.$$

First, a few words of warning: you are about to be beset by our only request in this book for you, the reader, to provide a real mathematical proof. It is elementary and short, and we believe in you; we believe you can pull it off. So please give it a try. It is also subtle, so if you have trouble with it, move on, but don't give up. If necessary, come back to it from time to time (here's a hint: the proof involves nothing more than being careful with the axiom of extension). We have little doubt that you will eventually get it.

And it is important to get it, because algebra in general and data algebra in particular is, well, couplets galore!

Taking Stock

The following bullet points summarize this chapter:

- Relations (between elements of sets) rather than functions (of the form $y = f(x)$ as in numerical algebra) are the gold standard of data algebra.
- The Cartesian product of A and B is defined by:

  $$A \times B := \{\{a\}, \{a, b\}\} \in \{P^2(A \cup B): a \in A \text{ and } b \in B\}.$$

- An element, $\{\{a\}, \{a, b\}\}$ of a Cartesian product is referred to as a couplet.
- The Cartesian product creates a set of couplets that can be illustrated in a tabular form. Couplets can represent Cartesian coordinates: points on a graph of a function, such as $y = x^2$.
- The Cartesian product "x" is neither associative nor commutative. However, it distributes over the union operator "U", as follows:

  $$(A \cup B) \times (A \cup B) = (A \times A) \cup (A \times B) \cup (B \times A) \cup (B \times B).$$

- Using the Cartesian product, $\mathbb{N} \times \mathbb{N}$, we can ascend from the set of natural numbers, $\mathbb{N}$, to the set of integers $\mathbb{Z}$. An involved explanation of how this works is provided, because we use a similar process of lifting to create algebras of data.
- We use the Kuratowski definition $A = \{\{a\}, \{a, b\}\}$ to define a couplet. In the sequel we discover that a "genuine" ordered pair is a certain set of two couplets and that an ordered n-tuple is a certain set of n couplets.
- Additionally, we introduce new notation (an exponential form) a^b to denote a couplet. This notation facilitates the eventual composition of couplets, and the so-called relations and clans they spawn, while keeping us from getting lost in forests of set braces.

Chapter 5: The Lion Sleeps Tonight

Go down deep enough into anything and you will find mathematics.

~ Dean Schlicter

CONSIDER the statement:

In the jungle, the mighty jungle, the lion sleeps tonight.

You probably recognize this as a line from a decades-old pop song. However, is it data? Or to declare our focus of interest and be more precise: is it data as far as a computer is concerned? We cannot accurately answer such a question without first answering the more precise question:

What is the fundamental unit of data?

When we hear the word "data" it is more likely to conjure up the idea of information stored in a spreadsheet, or a computer file or database than a sentence from a book or a line from a song. When we start to investigate this question, we quickly realize that it is not such an easy question to answer.

The algebraic necessity

Mathematically, we resolved the question of what data is, in a very general sense, by observing the fact that: data is a collection of "datums." This led us to set theory and thus to defining the rules we agreed to play by – the axioms of ZFC set theory.

But on its own, this does not provide us with a means of applying this to the data stored in a computer – and not just the data currently stored in computers, but any data that anyone may in the future decide to store in a computer.

It is an algebraic necessity that we start out with a sound definition of the fundamental unit of data which must apply even to data that may never be put into a computer. Consequently, it needs to be a very accurate definition and also one that is grounded in the world that we inhabit as human beings.

We might reason as follows: data defines our existence as living organisms, as human beings. Our means of understanding the world comes from information gathered by our six senses: touch, taste, smell, hearing, sight and the sense of motion and balance (proprioception). Our brain ingests a variety of such sensory impulses from our nervous system, sometimes combining one sense with another, and constructs a representation of the exterior world. We do this from before the moment of our birth, and we continue to do it until our last breath. So surely this representation of the world is a collection of data. Perhaps data is "the representation of anything we can perceive."

This is, in our view, a reasonable starting point, but it is not good enough for two reasons:

1. It ignores our ability to conceptualize.
2. It is subjective.

First we need to allow for the fact that we have the mental capacity to conceptualize, to represent concepts and even physical things that do not yet exist in reality. This too might be regarded as data, especially as such mental models may eventually become real-world data. The whole of mathematics is a collection of such mental models that have been committed to paper and shared extensively. Indeed, we could say the same about many conceptual models that inhabit the broad field of human intellectual endeavor.

The second point is much more troublesome for our initial stab at a definition of data. It is a philosophical and psychological cliché that when two people claim to see the color red, there is no guarantee that their experience is identical. And the existence of color blindness demonstrates that, in some circumstances at least, it definitely is not. Even if we could place electronic sensors in the human brain to record and identify signals emerging from the nervous system and were thus able to prove that most people receive identical signals at the interface to their brain, we cannot know for sure that their interpretation of them will be the same.

This means that we must go to objective sources, directly to the physical activity of man, in order to define what data is.

Human information

First consider text. For millennia humanity has communicated via the written word, chiseled into stone, or painted onto walls or written on papyrus or vellum or paper. The goal was to present data for others to read, whether they were contemporaries or future generations. We can thus examine data from this perspective.

Text, whether composed of alphabetic letters or glyphs, consists of words arranged into sentences. Numbers are the same and can be viewed as words of a kind created from a special alphabet (digits). So the basic building blocks of text are a set of symbols.

The first question then is whether these building blocks can be thought of as data. Although the symbols do not necessarily have explicit meaning in isolation – sometimes they do, sometimes they don't – when used together they impart meaning. So we suspect that they may constitute data, since they are components of data. However, with words, the units of text that the symbols create, there is much less doubt. Words have meaning.

For the sake of exactness we need to consider all arrangements of symbols that have meaning when taken together. So this includes numeric units of meaning, such as 3.14159, as well as verbal units of meaning, such as "lion" and "jungle." Words have meaning and numbers have meaning. As we construct sentences we create more sophisticated and more detailed meaning.

So does this provide a basis for us to define what we mean by data?

It doesn't.

The demands of objectivity

The problem is this. Words and numbers constructed in this fashion require a human being to decipher their meaning. If you ask several people what the word "lion" means, you will quickly realize that there is no reliable consensus as to its meaning.

The answer you get from any individual will vary according to their experience which, in respect to the word "lion," consists of a collcction of memories, some of which they will associate to when you ask the question. Their answer will almost certainly not be identical with a dictionary definition.

Look the word up in the *New Oxford American Dictionary*, for example, and it will suggest that a lion is "a large tawny-colored cat that lives in prides, found in Africa and northwestern India. The male has a flowing shaggy mane and takes little part in hunting, which is done cooperatively by the females."

The fundamental problem here is that the meaning people infer from "lion" varies according to their individual experience. Written down, the word "lion" is simply four alphabetic symbols in a row, pronounced in a particular fashion when verbalized. Consult a dictionary, and you will encounter a reasonable definition perhaps, yet even there you will not encounter agreement between dictionaries as to what should be included in the definition.

For contrast, consider the ***Merriam-Webster*** definition of "lion" (from merriam-webster.com) which states that a lion is "a large heavily built social cat (Panthera leo) of open or rocky areas chiefly of sub-Saharan Africa though once widely distributed throughout Africa and southern Asia that has a tawny body with a tufted tail and a shaggy blackish or dark brown mane in the male."

And it is only fair to point out that ***New Oxford American Dictionary*** also adds the following alternative meanings or usages:

- (the Lion) the zodiacal sign or constellation Leo.
- a brave or strong person.
- an influential or celebrated person: a literary lion.
- the lion as an emblem (e.g., of English or Scottish royalty) or as a charge in heraldry.

The *Merriam-Webster* alternative meanings or usages are not identical. They are:

- any of several large wildcats; especially: cougar.
- LEO (this, incidentally, is a link to the definition of LEO the constellation).
- a person felt to resemble a lion (as in courage or ferocity).
- a person of outstanding interest or importance <a literary lion>.

When the word "lion" comes up in conversation or is written on a page, any one of those meanings – some of which are metaphorical – may be intended. When we read the word, usually without any conscious effort, we select the intended meaning. We deduce its meaning from the context. So when we read the words:

In the jungle, the mighty jungle, the lion sleeps tonight,

we do not suppose that the constellation Leo, or some literary lion, or some brave or strong individual has nodded off in the jungle. We are pretty certain that Panthera leo himself has decided to call it a night.

This is all very well for human beings, but it creates a problem for computers. Computers can only make the estimations of meaning we have described here if they have similar mechanisms for determining meaning. And usually they do not have any such capabilities, although it is worth noting here that their semantic and reasoning skills are currently improving by leaps and bounds.

Consider something quite simple, such as an instant message from one person to another, as follows: "meet u after work 4 drink." Even someone who has never previously encountered the abbreviation "u" for "you" and "4" for "for" would be able to work out the meaning. They would deduce it. A computer, however, would be unlikely to have a basis for making such a deduction. Computers require data to be rigorously defined, to be objective. And so does an algebra of data.

The basic unit of data

We already know a great deal about what works as data from the way we use computers. We do not put information into a computer without representing it in a way that the computer can decipher and giving it sufficient context to allow us to do something useful with it. We know what we have found to be practical.

At the base level, we store information as encoded bits. The natural unit of computer data is called a "word." A specific definition for a word is required by the computer processor, since the instructions that the processor can execute only process data in units of words. In practice word sizes of 8, 16, 24, 32, 48 and 64 bits have been tried at various times and nowadays most computers use 32 bit or 64 bit words.

In general, although not always, the word determines the address size for addressing data held in memory. Machine instructions, the basic instructions executed by the processor, will refer to (point at) addresses of data held, and the data is fetched into the processor so that the instructions can be executed.

At the very basic level, ignoring a good deal of the nuances of computer architecture, we can think of computer programs as a series of instructions that specify addresses of data. When these instructions are executed by the processor, data is fetched, processed and the result is placed somewhere in memory. The instruction set that the processor executes is itself a set of binary codes, and the data contained in the words stored in computer memory is also binary. At this level, we could say that there are just two kinds of data: instructions stored in words and data stored in words.

At a slightly higher level the computer program distinguishes between different types of data. Exactly which data types are accepted by a computer program depends on the programming language. The following comprise the most common data types:

- The Boolean data type. This either has the value "True" (usually represented by a single bit with value 1) or "False" (represented by a single bit with value 0).
- The integer data type. This is an integer. Depending upon computer language and architecture there can be short integers, long integers and even long long integers, and they may be signed (meaning that they include the numeric sign, + or -). An unsigned long long integer can take any value from 0 up to 18,446,744,073,709,551,615.
- The floating point data type. This numeric form allows computers to represent and compute real numbers. They are called floating point because the number's decimal point can "float," so that very large numbers or very small numbers can be represented even if the representation is approximate to some degree. The number has two parts called the significand (an integer) and the exponent (also an integer). The number $3.14159 = 314159 \times 10^{-5}$ is thus stored as 314159 (the significand) along with -5 (the exponent). The limit on the size of the number that can be represented depends on the number of bits used for the significand and the exponent. The "single" floating point usually occupies 32 bits with 23 for the significand and 8 for the exponent and 1 for the sign. The double is 64 bits (52, 11, 1), and the quad 128 bits (112, 15, 1).

- The char data type. This is one character and is best thought of as a single keyboard character since the value usually corresponds to something that is visible on the computer keyboard.

That's all the basic data types, although most programming languages offer a richer capability than that, as do databases. The char naturally extends to become a string of characters, usually with a termination character to denote the end of the string. Date and date-time may have a dedicated data type. Arrays and lists are often supported by programming languages. There are no explicit data types for images or vector graphics or sound or video at the language level, but there are agreed standards (often several) for the representation of such data.

The point here is that values are not stored within a computer without there being some indication of what that specific value means. No matter where data values are stored, it is impossible to process them meaningfully without knowing what they represent.

In formulating the algebra of data, the couplet was chosen as the fundamental unit of data. Our discussion of the basic data types of a computer indicates that there is strong circumstantial evidence for suspecting that a couplet is the correct set-theoretical form. The bare metal computer is "objective" in the sense that it has no in-built mechanism for processing data that is not described to it explicitly. Its processing is couplet-based.

So a couplet appears to be a respectable starting point, as a "unit of meaning." If we put two values together in a couplet, for example, 9^{lions} (yes, we are using use the couplet notation from the previous chapter) we can think of each as qualifying or providing context to the other. That set might, for example, be a count of lions in a particular location. It is possible, of course, to randomly pick any two units of meaning and create a couplet, say, $telephone^{liquid}$, where the two values cannot meaningfully provide context for each other (except in a Salvador Dalí painting). This just indicates that some couplets are not particularly useful.

At the logical level

We can think of the computer as having two levels: a logical level where it represents data in a way that is convenient for people to use directly and a physical level where it stores data in a form that is convenient for its own needs. Any algebra of data worth our attention must be able to represent and operate on data at both levels and pass data from one level to the other.

So let us now consider the statement:

In the jungle, the mighty jungle, the lion sleeps tonight.

Is this data?

This certainly seems to be data that people process naturally. And without defining precisely what we mean by couplet mathematically, we can certainly represent this statement as a couplet in the following way:

$\textit{In the jungle, the mighty jungle, the lion sleeps tonight.}^{lyrics}$.

By doing this we have designated the statement as "lyrics." That's one way of thinking about it. However, you will no doubt have noted that this is indeed a sentence, and as such, we could, if we wished, express each word of it as a couplet grammatically. So, it may occur to you that we can represent it as the couplet S^s, where S = {lots of couplets} and s = lyrics:

$$\{In^{preposition}, the^{definite\text{-}article}, jungle^{noun}, ,^{comma}, the^{definite\text{-}article}, mighty^{adjective}, jungle^{noun}, ,^{comma}, the^{definite\text{-}article}, lion^{noun}, sleeps^{verb}, tonight^{adverb}\}^{lyrics}.$$

By the way: yes indeed, a set can be an element of a couplet (sets can be elements of sets). If we remove repetitive elements, we get:

$$\{In^{preposition}, the^{definite\text{-}article}, jungle^{noun}, ,^{comma}, mighty^{adjective}, lion^{noun}, sleeps^{verb}, tonight^{adverb}\}^{lyrics}.$$

If we wanted to preserve all the words in the correct order, we'd have to include order preserving information. This would give us:

$$\{\{In^{preposition\}}\}^{1}, \{the^{definite\text{-}article}\}^{2}, jungle^{noun}\}^{3}, \{,^{comma}\}^{4}, \{the^{definite\text{-}article}\}^{5}, \{mighty^{adjective}\}^{6}, \{jungle^{noun}\}^{7}, \{,^{comma}\}^{8}, \{the^{definite\text{-}article}\}^{9}, \{lion^{noun}\}^{10}, \{sleeps^{verb}\}^{11}, \{tonight^{adverb}\}^{12}\}^{lyrics}.$$

And this, too, is a couplet. But it is couplets within couplets within a set that's a component of a couplet. We imposed order on the set so that we did not lose meaning. And, of course, we can descend to a further level of detail if we wish to break up the individual words into their constituent letters. We might decide to transform every string that is noun and is recorded as the couplet $\{string^{noun}\}$ into such a set of compound couplets, as follows:

$$\{\{j^{letter}\}^{1}, \{u^{letter}\}^{2}, \{n^{letter}\}^{3}, \{g^{letter}\}^{4}, \{l^{letter}\}^{5}, \{e^{letter}\}^{6}\}^{noun}.$$

Thus, if we were so inclined, we could express the statement as a very complex structure, breaking it down into every single symbol it contains. And, by the way, to store this simple sentence in a computer we do indeed break it down to that level in some way – even though we currently do not do so in a respectable algebraic manner.

We are obliged to point out that we have made some assumptions about this sentence. We have, for example, assumed that the statement itself is just lyrics and not part of a wider context. Perhaps, in order to express the full meaning of what it represents, the data needs to be married with the musical notes that accompany it. There is nothing explicit about the original statement that insists it is not music. So, perhaps the data we should be storing is this:

Or perhaps we should only be storing a sound recording of that line of the song. And since it is a song, and thus covered by copyright, perhaps we need to investigate whether we have the legal right to store this in any form

whatever. Perhaps we do not.

Taking another tack, we are actually presuming that this data refers to the lyrics of a song. Perhaps this is not the case. Perhaps it really is just a factual statement intended to convey meaning, in which case we can represent it as a couplet in the following way:

Tonight the lion sleeps in the mighty jungle^{fact}.

Notice here that the order of words may no longer be so important, since several different versions of this sentence will be comprehensible to any reader (The lion sleeps tonight in the mighty jungle; The lion sleeps in the mighty jungle tonight, and so on).

But sadly, with this example, we are unlikely to want to classify it as fact, since lions are never found in jungles. Their habitat is the savannah, where they hunt buffalo, zebras and wildebeest, and where, sometime after nightfall, they go to sleep.

Taking yet another tack, we note that we are making definite assumptions about the individual to whom the data may be served up. The data consumer might be illiterate, or even if literate, the English language may be completely foreign to him or her. It is even possible that the data consumer does not recognize the English alphabet, and thus if they encounter the sentence, they may not even be certain that it is genuine text in some foreign language.

Consider this:

Do you know whether this is genuine text or not? Do you even know

է ջունգլիներում հզոր ջունգլիներում առյուծը քնում երեկո

whether you should read it from right to left or left to right? You probably don't. In fact, it should be read from left to right. It is Armenian script. Roughly translated into English, it reads:

In the jungle, the mighty jungle, the lion sleeps tonight.

The point is that information may have structure in the sense that, if served up to the right person, meaning can be gleaned from the information, but the computer itself does not explicitly know this unless it is given some indication that this is so.

In connection with this, we note that the term "unstructured data" is frequently used in IT. However, it is poorly chosen terminology because it implies, when interpreted literally, that there can be some data which has no structure whatsoever. In fact, the meaning of "unstructured," according to Wikipedia, is:

> *Unstructured data (or unstructured information) refers to information that either does not have a pre-defined data model or is not organized in a pre-defined manner. Unstructured information is typically text-heavy, but may contain data such as dates, numbers, and facts as well.*

In other words "unstructured data" can have implicit structure but has no explicit structure. Our example is, by this definition, "unstructured" data. It is certainly the case that we could feed the data into a computer and have it recognize the input as text. However, it could not know for sure that it was text; it could only know that it matched the pattern of text. Nevertheless, if that is what we wanted the computer to do, then we would have applied a function, let us say, *f* with the following effect:

f (*In the jungle, the mighty jungle, the lion sleeps tonight.*^{input}) :=
In the jungle, the mighty jungle, the lion sleeps tonight.^{text}.

It is difficult to imagine any data that has no structure. Even if we think of the various radio telescopes involved in the SETI project, which listen to and record radio waves from everywhere in the sky, hoping to find evidence of extraterrestrial life, we cannot know for sure that what they have recorded is simply random noise. All we know is that no one who has analyzed such signals has yet been able to detect structure in them.

For all we know, some of these signals have a structure that we have failed to recognize. Perhaps, buried somewhere in that extensive record of extraterrestrial radio signals, is the plaintive sound of some alien voice singing *The Lion Sleeps Tonight.*

The subjectivity of man

Hopefully, we have said enough to demonstrate that although the world of man has aspects of objectivity – people tend to understand each other when they speak the same language – the world of the computer is more demanding. In the world of man, our simple example, *In the jungle, the mighty jungle, the lion sleeps tonight.*, qualifies as data among a subgroup of humanity that recognizes the English language. However, it only does so because we have inner intellectual mechanisms which can parse the data and categorize it accordingly.

The computer, completely lacking any such mechanism, needs to have the context of that string of characters specified. As such, we have no option but to declare that *In the jungle, the mighty jungle, the lion sleeps tonight.* is not data as far as a computer is concerned. However, if we turn it into a couplet, as follows:

In the jungle, the mighty jungle, the lion sleeps tonight.^{string}.

then the computer can treat it as data at the physical level. We can write programs, which manipulate strings and feed such data to them. If we want to define this within the computer in a way that might be useful to people, we may prefer that couplet to be defined as follows:

In the jungle, the mighty jungle, the lion sleeps tonight.^{lyric}.

or possibly, we might be satisfied with:

In the jungle, the mighty jungle, the lion sleeps tonight.^{text}.

All of this suggests that the objective unit of data genuinely is the couplet, and hence couplets are an excellent place to begin in an effort to construct an algebra of data.

Taking Stock

This chapter dealt with the question: What is data? The following bullet points provide a summary:

- The inner (psychological) world of man is demonstrably subjective.
- The outer consensus world of man is also demonstrably subjective. Where it is able to achieve objectivity, person to person, it does so through context.
- At the basic physical level within a computer, data is represented using couplets. One part of the couplet specifies value, the other part specifies usage.
- Couplets can also be used for representing data at the logical level, in terms that people can understand.
- The algebra of data takes the couplet to be the fundamental unit of data.

Chapter 6: Yin and Yang

How can it be that mathematics, a product of human thought independent of experience, is so admirably adapted to the objects of reality?

~ Albert Einstein

IF YOU HAVE FOLLOWED our trail of breadcrumbs into the not-so-fearsome forest of data algebra, you will now be familiar with power sets. You will have understood their importance: that you can build a tower from such sets and that you can climb energetically from one level to another, lifting existing operators and creating new operators as you go, as dictated by opportunity and need. You will have refamiliarized yourself with unions, intersections, complementation and possibly even with Cartesian products, if you previously met them at some point of your education.

You have just been introduced to the idea that the couplet, a^b, is the true basic unit of data and, to carve this idea in stone, you've been led through a discussion of what human beings happily accept as data and what computers are constrained to accept as data. The main point to emerge from this is that computers are an aggregation of dumb electronic circuits that need to be precisely informed about data, while human beings, blessed with an associative memory and some ability to reason, can manage with less precision. As the algebra of data is built for the computer's world, it conforms to the limits of the computer's world.

Despite the fact that we just spent a chapter discussing the validity and importance of the couplet, we have more to say. Basic units are indeed important...

A little more about couplets

We can commence with the fact that all civilizations develop or inherit units of measure. Such units, whether they apply to time, length, angle, weight, liquid volume or money, become the foundation of many activities, and without these units of measure, the activities would not be possible. If, for example, you were an apothecary in Great Britain in any century prior to the 20th, you would have known that there are: 20 grains to a scruple, 3 scruples to a drachm, 8 drachms to an ounce and 12 ounces to a pound. You would also have had a set of weights that allowed you to weigh out such quantities, so that you could properly cost raw materials and price the various remedies you sold. Nowadays, of course, we have streamlined units based on decimal multiples. With only a few exceptions, we measure distances with meters, weights with grams, liquid volume with liters and so on.

So when we encounter a number, it is usually accompanied by a unit, suggesting very strongly that the numbers of the real world are really couplets of the form:

$$n^{unit}.$$

This is also true of the so-called pure numbers of theoretical mathematics. Those devil-may-care mathematicians have been happily writing numbers such as 1, 1.618, 3.14159 and so on as if they weren't couplets, and they are. And so are all those variables, x, y, and z, that star in their magic algebraic equations; couplets, every one of them. Depending on what they are, we can formally write them as one of the following:

$$n^{natural}, i^{integer}, q^{rational}, r^{real}, c^{complex}.$$

But of course, mathematicians finesse all of that confidently and safely assuming that the lowliest number they deal with lives somewhere in that notationally misleading tower:

$$\mathbb{N} \subset \mathbb{Z} \subset \mathbb{Q} \subset \mathbb{R} \subset \mathbb{C}$$

Couplets (Γ)

Couplets (Γ) is an algebra. Once we have told you all there is to know about it, you may conclude that it is meager and unexciting; a couple of operators and a collection of data to get algebraic with – so what? However, this is a rags to riches story, and the rags of the couplet algebra will look more and more like royal robes as we climb the algebraic tower. So wait for that with bated breath.

In the meantime, as we're talking about Couplets (Γ), let's talk about the genesis set. The couplets have to come from somewhere, and the genesis set Γ is the bag of things whence they emerge. So what's in the bag? Officially and pragmatically, it's the "local" universe of "stuff," which is sufficient to generate all the couplets we wish to consider and manipulate. And just to emphasize the point, it is finite, even if large.

Γ x Γ		Γ		
		leather	*red*	*yellow*
Γ	*leather*	$leather^{leather}$	$leather^{red}$	$leather^{yellow}$
	red	$red^{leather}$	red^{red}	red^{yellow}
	yellow	$yellow^{leather}$	$yellow^{red}$	$yellow^{yellow}$

To get the couplets we want, we manufacture them using a Cartesian product. Imagine if there were only three things in the bag Γ: red, yellow, leather.

Not much content, but we would still be able to construct 9 couplets as follows:

And while we might not be able to do much with those couplets, we can at least list the well-known tongue twister:

$$red^{leather}, yellow^{leather}, red^{leather}, yellow^{leather}.$$

If only we had some functions...

The functions: yin and yang

We will, at times, want to get our hands on the one or the other of the two components of a couplet, so we need a pair of functions to extract them. This pair of functions are named *yin* and *yang*, terms borrowed from Chinese philosophy, where *yin* and *yang* are complements rather than direct opposites. These words were not chosen by accident, but selected because they carry no implication of order.

So let there be two sets, *A* and *B*, with:

$$a^b = \{\{a\},\{a, b\}\} \in A \times B.$$

Then:

$$yin\ (a^b) = a \text{ and } yang\ (a^b) = b.$$

And now, without missing a beat, let's talk about the transposition operator. It should come as no surprise that we can transpose couplets. Recall back in Chapter 4, when we climbed the numeric tower, you will remember transposition,↔, to be a unary operator, that "switches things around." This is how it works on couplets:

$$b^a = a^b$$

So, as you can immediately deduce, the transpose of $red^{leather}$ is $leather^{red}$, that the transpose of a transpose is, as always, what you started with and the transpose of a couplet like $yellow^{yellow}$ is just the same, $yellow^{yellow}$.

If we now introduce *yin* and *yang*, and we can declare that:

$yin\ (red^{leather}) = red,$

$yang\ (red^{leather}) = leather.$

And if you transpose the $red^{leather}$ to become $leather^{red}$ then the *yin* and *yang* results switch around so that:

$yin\ (leather^{red}) = leather,$

yang $(\text{leather}^{\text{red}}) = \text{red}.$

As you will realize later, given the transpose operator and the *yin* and *yang* functions, we can get at any part of a couplet whenever we want.

Composition

We promised you two operators for our modest Couplets (Γ) algebra, so let's deliver on this promise. The second operator is a binary operator, and it's useful, and it's called *composition*. When the composition operator, "∘" composes two couplets, it behaves like this:

$$a^b \circ b^c = a^c.$$

To precisely define composition, we express it as follows:

$$a^b \circ d^c = a^c \text{ if, and only if, } b = d.$$

You may be wondering, "Well, what the hell happens if b is not equal to d?" The answer is, "In that case, there is no defined result." It's what is called a partial operator; sometimes it gives a meaningful answer, and sometimes it just doesn't.

Consider this for example:

$$red^{leather} \circ leather^{yellow} = red^{yellow}.$$

whereas:

$$red^{leather} \circ red^{yellow} \text{ is undefined.}$$

That's the downside of composition in Couplets (Γ) but there is also an upside. Composition is associative!!

OK, we realize that you are not going to leap out of your chair and cheer just because of that, but mathematically, it's a real positive. Composition is associative:

$$a^b \circ (b^c \circ c^d) = a^b \circ b^d = a^d = a^c \circ c^d = (a^b \circ b^c) \circ c^d,$$

which is definitely associativity.

Or we can write:

$$red^{leather} \circ leather^{yellow} \circ yellow^{red}$$

and it doesn't matter which composition we work out first; we'll end up with the answer red^{red}.

The transposition of the composition

Now for something unexpected. We'd really like it if the transposition of a composition was the composition of the transpositions. Apart from anything else, we'd be able to blurt out our *flarn-of-the-clarp* mantra. But it isn't. The transposition of the composition is actually the composition of the transposition in the reverse order!

$$a^b \circ b^c = a^c = c^a = c^b \circ b^a = b^c \circ a^b$$

And, as long as you remember this curious backflip when you're wrestling with data, it will serve you well. (*Hint*: we mean it.)

Composition within the computer

Composition may not seem particularly useful at first, but it is the kind of operation that happens within a computer time and again – it just doesn't happen mathematically (yet). We know for example that a data item, when defined in a computer program, may be defined as:

$$Int^{age}.$$

Thus, a metadata tag (age) is bound to a data type (Int). The choice of data type was made most likely by the programmer. But when the computer comes to process the data item, the processor core will specify a memory address where that Int is to be found. Let us imagine that this address is the 64 bit value EG1A037F. At that level of activity, the computer is not concerned with what the integer residing at that location represents. It will simply think in terms of:

$$E61A037F^{Int}$$

when it fetches the data for processing. And if it uses the data in some calculation or even simply moves it somewhere else, it will be handling the integer's value. Let's say the value is 35. Then the computer will have carried out the composition:

$$35^{EG1A037F} \circ EG1A037F^{Int} = 35^{Int}.$$

And if it later uses the metadata tag, age, to describe the data when it displays it in a window, it may carry out the composition:

$$35^{Int} \circ Int^{age} = 35^{age}.$$

In general, we can express an item of computer data with the couplet:

$$((value^{location})^{type})^{tag}.$$

As we have not shown you a compound couplet like that before, it may not have occurred to you that we could create such a thing. However, the truth is we can include anything we choose in a genesis set, including a compound couplet such as $(value^{location})^{type}$. The *yin* of a couplet can be a couplet and so can the *yang*. And we can apply our oriental functions to get at the bits we want.

So, for example:

$$yin\ (yin\ ((value^{location})^{type})^{tag}) = value^{location},$$

$$yin\ (yin\ (yin\ ((value^{location})^{type})^{tag})) = value,$$

$$yang\ ((value^{location})^{type})^{tag} = tag,$$

$$yang\ (yin\ ((value^{location})^{type})^{tag}) = type,$$

$$yang\ (yin\ (yin\ ((value^{location})^{type})^{tag})) = location.$$

Also, introducing some composition:

$$yin\ (value^{location} \circ location^{type}) = value,$$

$$yang\ (value^{location} \circ location^{type}) = type.$$

And by the way, we could have driven you half insane by throwing the transpose operator into the various equations shown above and making the data do somersaults and backflips, but we judged it unwise.

Further extensions to couplets

There isn't really any limit to how complex a couplet can become, because a couplet can always be the *yin* of a couplet, or the *yin* of the *yin* of the *yin* of a couplet, and so on.

Consider a lion, whose name is Ralph. The following couplet is quite valid, and it's meaningful:

$$((((((Ralph^{lion})^{panthera})^{felidae})^{carnivora})^{mammalia})^{chordata})^{animalia}.$$

There is quite a lot of knowledge stored implicitly in this couplet, since, aside from the fact that Ralph happens to be of the species "lion," that species is part of the genus "panthera," which in turn is of the family "felidae," which is of the order "carnivora" within the class "mammalia" inside the phylum "chordata," all sitting inside the kingdom "animalia."

This demonstrates the possibility of adding qualification after qualification to a couplet. However, if you prefer, you can completely reverse it from beginning to end and represent the same data as:

$$((((((animalia^{chordata})^{mammalia})^{carnivora})^{felidae})^{panthera})^{lion})^{Ralph}.$$

As already noted once or twice, a couplet does not imply any ordering of any kind. So both representations are fine, although one may be preferable to the other depending on how you intend to use the data. If you examine either of them, you may think that some additional information needs to be explicitly included such as words like "species," "genus," etc. We could capture that information using a set of couplets. One way to do that would be:

$$\{Ralph^{lion}, lion^{panthera}, panthera^{felidae}, felidae^{carnivora}, carnivora^{mammalia}, mammalia^{chordata}, chordata^{animalia}, species^{lion}, genus^{panthera}, family^{felidae}, order^{carnivora}, class^{mammalia}, phylum^{chordata}, kingdom^{animalia}\}.$$

But we cannot do that in our lowly Couplets(Γ) algebra; we will have to lift it up the tower to do that. However before we do, let's declare a signature for Couplets(Γ).

The signature of Couplets (Γ)

While couplets are based on a *genesis set* Γ, the *ground set* from which the couplets emerge is the Cartesian product of Γ with itself, Γ x Γ. We can thus propose that the signature of Couplets (Γ) is:

$$[\Gamma \times \Gamma, \{\circ\}, \{\leftrightarrow\}].$$

In plain English, this is saying that the algebra Couplets (Γ) involves couplets created by the Cartesian product of Γ with itself and includes two operators, the partial binary composition operator and the unary transposition operator. And, by the way, that binary composition operator is not just partial, it also has no identity.

We have seen something similar to this with the natural numbers ℕ and its associated algebra being limited but its capability being increased by lifting it to the integers ℤ. So witness what happens when we ascend the algebraic tower by lifting Couplets (Γ) to Relations(Γ).

Relations (Γ)

First things first; a relation is a set of couplets. If we have some couplets, say, $Thor^{hammer}$, $Hulk^{anger}$, $Iron\ Man^{suit}$, and we wish to relate them to each other in a set, we can wrap them up in braces like this:

$$\{Thor^{hammer}, Hulk^{anger}, Iron\ Man^{suit}\},$$

and we have created a relation. Couplets, as you well know (we hope) are created by Cartesian products. So the existence of those couplets implies the existence of corresponding sets, say:

A set of comic book superheroes, $\{Thor, Hulk, Iron\ Man,... \} \subset \Gamma$.
A set of comic book superhero powers, $\{anger, suit, hammer,... \} \subset \Gamma$.
A Cartesian product, $\Gamma \times \Gamma = \{Thor^{hammer}, Hulk^{anger}, Iron\ Man^{suit},... \}$.

Relations(Γ) is the algebra of all sets R, where $R \in P(\Gamma \times \Gamma)$. In particular,

$$\{Thor^{hammer}, Hulk^{anger}, Iron\ Man^{suit}\} \in P(\Gamma \times \Gamma).$$

Incidentally, you may be wondering why we chose to call a set of couplets a relation. Perhaps you suspect the word has been chosen at random, and we could just as easily have called it a quark or a boson. Or maybe you believe we are intent on irritating aficionados of relational theory who, when asked "What is a relation?" will often reply, "It's a table, of course."

In truth, the word "relation" wasn't so much chosen as it chose itself. A set of couplets was created for some reason that justified the assembly of its elements, and that reason, no matter what it was, is a relation. Even if you decide to randomly pick a set of apparently unrelated couplets, it is a relation, and the relation that holds the elements together is "the set of apparently unrelated couplets I chose to pick."

Come yin, come yang

So now that we have a new tier of the tower to explore, let's see what we can do with our beloved functions *yin* and *yang*. Luckily, there is an easy and obvious way to apply them to relations. Let's show and tell. The *yin* of a relation is the set of the *yins* of the elements of the relation, and naturally, the *yang* of a relation is the set of the *yangs* of the elements of the relation. So, consider:

$$R = \{Pinocchio^{puppet}, Miss\ Piggy^{muppet}, Popeye^{cartoon}\}.$$

Then:

$$yin\ R = \{Pinochio, Miss\ Piggy, Popeye\}$$

is the yin set of *R* and:

$$yang\ R = \{puppet, muppet, cartoon\}$$

is the yang set of *R*.

Also, conveniently and with little intellectual effort, we can lift the operation of transposition into the algebra of relations:

$$\overleftrightarrow{R} = \{\ b^a \in B \times A:\ \overleftrightarrow{b^a} \in A \times B\}.$$

In English: *The transpose of the relation R is the relation whose elements are the transposes of the elements of R.* It's one of those *flarn-of-the-clarp* situations. A simple cartoon example will make this even clearer:

$$T = \{Bugs^{Bunny}, Daffy^{Duck}, Wile\ E.^{Coyote}\},$$

$$\overleftrightarrow{T} = \{Bunny^{Bugs}, Duck^{Daffy}, Coyote^{Wile\ E.}\}.$$

Now that we have the defined the transpose of a relation, we can introduce some new and useful yin-yang terminology. So let's introduce the important concepts of yin-functional and yang-functional relations:

> If $a^b \in R$ and $a^c \in R$ imply that $b = c$ then *R* is said to be yin-functional.

As is visually apparent, our relation *T* is yin-functional.

We can concoct a similar definition for the meaning of yang-functional:

$\boldsymbol{R}$ is yang-functional if $\overleftrightarrow{\boldsymbol{R}}$ is yin-functional.

Take a look at *T* and you see also that *T* is yang-functional. And just to be sure that you've got this, we'll consider a different cartoon relation:

$$C = \{Donald^{Duck}, Daffy^{Duck}, Yogi^{Bear}\}.$$

This *C* is again yin-functional, but it is not yang-functional. And of course, if we consider its transpose, $\overleftrightarrow{C}$, it is yang-functional but not yin-functional.

The composition of relations

Remember how, when we raised $\mathbb{N}$ up the tower to $\mathbb{Z}$, we gained the unary operator – ? Well, when we raise the composition operator up from Couplets(Γ) to Relations(Γ), we fix the uncomfortable fact that compositions of couplets can be undefined. Neat, huh?

For couplets, we defined composition to be:

$$a^b \circ d^c = a^c \text{ if, and only if, } b = d$$

and when b was not equal to d, the outcome was undefined.

You may have wondered why we didn't just declare the outcome to be Ø. We would have liked to, but we couldn't bring ourselves to do it, because Ø is not an element of the Cartesian product $\Gamma \times \Gamma$. We would have violated our own rules, and the mathematical umpire would have penalized us and deducted any points we thought we might have scored.

With the composition of relations, we can have Ø be the result of the composition of two singleton sets of couplets, because relations are sets. Indeed we can define the outcome to be in the pattern of *the-flarn-of-the-clarp-is-the-clarp-of-the-flarns* or using real terminology, the composition of the relations is the relation of the (defined) compositions of couplets.

So consider this example:

$$\{a^b, b^c, c^d\} \circ \{b^c, d^e\} = \{a^c, c^e\}.$$

Mathematically, the only compositions of couplets that are defined are $a^b \circ b^c = a^c$ and $c^d \circ d^e = c^e$, so the result is $= \{a^c, c^e\}$.

And here's an example with real values, where we compose poets and painters, rather than poetry and paintings:

$$\{Blake^{Poet}, Dali^{Painter}, Byron^{Poet}\} \circ \{Painter^{Surrealist}, Poet^{Romantic}\}$$
$$= \{Blake^{Romantic}, Dali^{Surrealist}, Byron^{Romantic}\}.$$

To make it crystal clear, if we try the following composition,

$$Blake^{Poet} \circ Byron^{Poet}.$$

the result is undefined. However,

$$\{Blake^{Poet}\} \circ \{Byron^{Poet}\} = \emptyset$$

is slightly different, giving the empty set as the result. The point is that ascending the algebraic tower smooths the path nicely.

And, just in case you wondered about it, the composition of relations turns out to be associative, so for $A, B, C \in$ Relations(Γ):

$$A \circ (B \circ C) = (A \circ B) \circ C.$$

You might have expected this, because composition (when there were defined results) was associative in Couplets(Γ).

However, composition is not commutative – which is a damn shame, but what can you do? You were probably able to quickly work that out, assuming you remembered what "commutative" meant. If you didn't, we'll remind you: it means the result is the same whatever the order of the operands. So + is commutative because $a + b = b + a$, and ∘ is not because:

$$\{a^b, b^c\} \circ \{c^a\} = \{b^a\}, \text{ whereas } \{c^a\} \circ \{a^b, b^c\} = \{c^b\}$$

or alternatively:

$$\{Byron^{Poet}, Poet^{Mad\ man}\} \circ \{Mad\ man^{Byron}\} = \{Poet^{Byron}\},$$

whereas:

$$\{Mad\ man^{Byron}\} \circ \{Byron^{Poet}, Poet^{Mad\ man}\} = \{Mad\ man^{Poet}\}.$$

Practical application

We have described relations mathematically, so let's now discuss how they manifest themselves in the digital world of computers. Algebraically, many of the files held on a PC are relations.

For example, a photograph is usually stored as an image, maybe with some contextual data provided by the digital camera, such as when the photo was taken, where it was taken and what the camera settings were. In other words, it is a set of couplets. Digital photography standards, such as TIFF, define such relations explicitly.

Simple text files (.txt files) on a PC are relations. On the PC, they have the characteristic extension ".txt," and each line of text in the file, except the last, is terminated with the ASCII CR (carriage return) and LF (line feed). If we had a file called Example.txt with three lines of text, then the relation could be defined as:

$$E = \{txt^{filetype}, line1^{text}, line2^{text}, line3^{text}\}.$$

If we wished to treat the file algebraically, we would read it and transform it into a relation while we did so. In general, audio files and video files are handled in a similar way to .txt files, breaking up the data into chunks. And they, too, are relations, as are email files.

For all files that are commonly shared between different programs (text, images, graphics, video, music and so on) there are standard data formats invented by software companies or agreed upon by standards bodies. There have been efforts to formulate more general standards for this purpose, such as XML.

XML (it stands for the eXtensible Markup Language) is an open standard that provides a set of rules for encoding information (often documents) using tags. It is an open standard intended to be understood by humans and computers alike. Below is a very simple example:

```
<?xml version=“1.0” encoding=”UTF-8”?>
<note>
  <to> Reader</to>
  <from>Authors</from>
  <heading>Promotional Message</heading>
  <body>Data algebra is way cool!</body>
</note>.
```

With XML, it is standard for the first line to declare the XML version number and the encoding used. The rest of the information is data, delimited by tags, which can easily be translated into a relation called *Note*, as follows:

$$Note = \{Reader^{to}, Authors^{from}, Promotional\ Message^{heading}, Data\ algebra\ is\ way\ cool!^{body}\}.$$

If we wanted to select just the "to" and "body" from the relation we can simply construct another relation called *Select* as follows:

$$Select = \{to^{to}, body^{body}\}.$$

and then we could evaluate:

$$Note \circ Select = \{Reader^{to}, Data\ algebra\ is\ way\ cool!^{body}\}.$$

Let us now consider another common computer file, a CSV file. For this example, let's return to our set of superheroes *A* and their "powers" *B*. Each consists of just three elements, so that:

$$A = \{Thor,\ Hulk,\ Iron\ Man\}$$

and:

$$B = \{hammer,\ anger,\ suit\}.$$

We can take the Cartesian product *A* x *B* of these two sets. This would give us the set of 9 couplets displayed in the following array:

A x *B*		*B*		
		hammer	*anger*	*suit*
A	*Thor*	$Thor^{hammer}$	$Thor^{anger}$	$Thor^{suit}$
	Hulk	$Hulk^{hammer}$	$Hulk^{anger}$	$Hulk^{suit}$
	Iron Man	$Iron\ Man^{hammer}$	$Iron\ Man^{anger}$	$Iron\ Man^{suit}$

That's fine, but it gives us six couplets we are not really interested in. The only ones we are interested in are those where our superheroes are married with the source of their powers. So imagine that we were provided these two sets, *A* and *B*, in the form of a CSV file on a computer, as follows:

Thor, Hulk, Iron Man
"hammer", "anger", "suit".

We are familiar with what a CSV file is. So we know that there is an implicit order in the elements of the two records in such a file. When parsing the records, as we read them from disk, we might turn them into ordered 3-tuples as follows:

$$C = \{1^{Thor},\ 2^{Hulk},\ 3^{Iron\ Man}\}$$
$$D = \{1^{hammer},\ 2^{anger},\ 3^{suit}\}.$$

That would allow us to transpose *C* to become $\overleftrightarrow{C}$ and compose it with *D* to give $\overleftrightarrow{C} \circ D$, which would look like this:

$\overleftrightarrow{C} \circ D$		D		
		1^{hammer}	2^{anger}	3^{suit}
$\overleftrightarrow{C}$	$Thor^{1}$	$Thor^{hummer}$		
	$Hulk^{2}$		$Hulk^{anger}$	
	$Iron\ Man^{3}$			$Iron\ Man^{suit}$

and now we have what we were hoping for:

> Thor is united with his hammer,
> Hulk is married to his characteristic ill temper,
> and Iron Man is wearing his shiny yellow and red suit.

So, by composing two relations of cardinality 3, we end up with a relation of cardinality 3, that looks like this:

$$\{Thor^{hammer}, Hulk^{anger}, Iron\ Man^{suit}\}.$$

Diagonals

We need to clue you in about diagonals so you don't react adversely or even run off screaming when we tell you what the signature of Relations(Γ) is. And, to be honest, there is no need to run off screaming, because diagonals are fairly simple. Consider:

Γ x Γ		Γ			
		a	b	...	z
Γ	a	a^a	a^b	...	a^z
	b	b^a	b^b	...	b^z
	...	...	...	...	...
	z	z^a	z^b	...	z^z

This Cartesian product is the great and glorious couplet factory that manufactures every possible couplet of data that will ever find its way into a computer.

Now consider $\mathfrak{P}(\Gamma \times \Gamma)$. It is a huge mountain of sets containing all those couplets as sets, in every collection of couplets you can imagine. And every collection of couplets in $\mathfrak{P}(\Gamma \times \Gamma)$ is a relation, which means that the set of couplets which make up the main diagonal of the above display is also a relation.

We denote this relation as D_Γ, defining it as follows:

$$D_\Gamma := \{a^a \in \Gamma \times \Gamma : a \in \Gamma\},$$

which means that:

$$D_\Gamma = \{a^a, b^b, \dots, z^z\}.$$

Here we have to pedantically make the point that the table shown above is a convenient visual artifice that illustrates the elements of the diagonal in a convenient way for the reader. The importance of D_Γ is that it serves as the identity for the composition in Relations(Γ):

$$R \circ D_\Gamma = R = D_\Gamma \circ R.$$

It turns out that subsets of D_Γ are also very important in Relations(Γ). We hope you'll realize later when we start to manipulate data more aggressively than we have done so far.

We note, in passing, that:

$$yin\ (D_\Gamma) = yang\ (D_\Gamma) = \Gamma.$$

The signature of Relations(Γ)

The signature of Relations(Γ) is:

$$[\mathrm{P}(\Gamma \times \Gamma), \{[\circ, D_\Gamma]\}, \{\leftrightarrow\}].$$

This announces that all the relations deriving from the genesis set Γ are elements of the power set P(Γ x Γ) and that Relations(Γ) includes a binary operator ∘, whose identity is D_Γ and a unary operator ↔.

To be clear, the *genesis set* of Relations(Γ) is the same Γ that gave birth to Couplets(Γ), but its *ground set* is the glorious power set P(Γ x Γ), which has implications that you might not have yet realized. So to remind you about power sets...

If we have a set $A = \{a, b, c\}$, the power set P(A) is:

$$\{\emptyset, \{a\}, \{b\}, \{c\}, \{a, b\}, \{a, c\}, \{b, c\}, \{a, b, c\}\}.$$

This consists of every possible subset of A, including the set A itself and the empty set $\emptyset$. When we construct the power set of the Cartesian product $\Gamma \times \Gamma$, $P(\Gamma \times \Gamma)$, we get every possible set of couplets Γ can furnish. The elements of $\Gamma \times \Gamma$ are couplets, but the elements of $P(\Gamma \times \Gamma)$ are all sets of couplets. So they are relations.

What the signature of Relations(Γ) fails to mention is that $P(\Gamma \times \Gamma)$, by its very nature, supports a set-based Boolean algebra. So, in climbing the tower, we have not only taken two useful operators with us, we have purloined the set operators, union $\cup$, intersection $\cap$, complement $'$ and the subset relation $\subset$. They are all valid for Relations(Γ), but by mathematical convention, rather than through modesty, we do not draw attention to that in the signature.

A word about functions

If you hark back to Chapter 4, you may remember we warned you about functions. Very specifically, we warned you that all those lovely well-behaved $y = f(x)$ types of functions that you enjoyed in school, or possibly at university, are not a feature of data algebra. You are probably interested to know why. The answer is not complicated, but it is important.

Those $y = f(x)$ functions are just one example of a function. Mathematically, a function is something that, when provided with an input, gives you a unique output. So if we have $f(x) = 3x + 2$, and we provide the input $x = 5$, then the output will be 17. And because we can take values of x very, very close to each other, when we plot the output on paper with Cartesian coordinates, it will look like a straight line. And if the function was $f(x) = x^2$, the graph would look like a parabola. Numeric algebra, with its beautiful polynomial functions, can easily seduce you into believing that all mathematical life can be like that. It cannot.

We can also have a function f(person) = SSN, where SSN is a person's social security number. This doesn't conveniently translate into a graph with axes. Neither is there some known calculation or procedure can be applied in some fashion, where if you plug a person into the function it naturally calculates the SSN. All you have is a set of inputs (people) and a set of outputs (SSNs). In fact, what you have if you take the input and the output together is a set of couplets, each couplet of which is of the form: $person^{SSN}$.

So when you encounter other data algebra functions later in this book, please remember that.

Ordered n-tuples

We shall round up this chapter by considering "the mysterious case of the missing Cartesian product." We noted earlier – you may have missed it, but believe me, we did draw attention to it – that Cartesian products of the form $A \times B \times C$ are not kosher. In fact, to be precise, we wrote "$A \times B \times C$ is notational nonsense unless $A = \emptyset$ or $B = \emptyset$ or $C = \emptyset$."

That's all very well, but what if we want to create some bona fide 3-tuples? Let's examine a set of ordered 3-tuples that we can construct from the set U. We start with an ordered set of cardinality 3, the set of integers {1, 2, 3}. An ordered 3-tuple is an element of the Cartesian product:

$$\mathrm{P}(\{1, 2, 3\} \times U).$$

A power set of a Cartesian product is not the simplest set in the world. It is peppered with many couplets that we have very little interest in here. Nevertheless, it will contain some sets of the form $\{1^a, 2^b, 3^c\}$ where $a, b, c \in U$. That is, it will contain some sets of three couplets that are ordered 3-tuples.

Now let $A \cup B \cup C \cup U$. We can then designate $A \times B \times C$ to be:

$$\{\{1^a, 2^b, 3^c\} \in \mathrm{P}(\{1, 2, 3\} \times U): a \in A,\ b \in B \text{ and } c \in C\}.$$

This resolves the mysterious case of the missing Cartesian product, $A \times B \times C$, although to be honest, you probably had no idea that there was a mysterious case in progress, even though there was. But never mind, we have a capability here to create ordered pairs, ordered triples and ordered n-tuples – all of which are sets of couplets. The general case is:

If $A \cup B \cup C \cup ... \cup N \cup U$, we can designate $A \times B \times C \times ... \times N$ to be

$$\{\{1^a, 2^b, 3^c ... , n^z\} \in \mathrm{P}(\{1, 2, 3 ..., n\} \times U): a \in A,\ b \in B,\ c \in C,\ ...,\ z \in N\}.$$

And, please note, those 1^a, 2^b, 3^c... etc. are elements of a set that we happen to have written in a specific order but could be written in any order. It is the yin set of the n-tuple, $\{1, 2, 3 ..., n\}$, that imposes an order, and it does so because it is an initial segment of the natural numbers $\mathbb{N}$, not because of any apparent arrangement of the list.

If you have a sharp mathematical eye, you may have noticed that we actually introduced ordered n-tuples on page 64, without deigning to explain them, when we chose to enforce an order on the following collection of text:

$$\{\{In^{preposition\}}\}^{1}, \{the^{definite\text{-}article}\}^{2}, jungle^{noun}\}^{3}, \{,^{comma}\}^{4}, \{the^{definite\text{-}article}\}^{5}, \{mighty^{adjective}\}^{6}, \{jungle^{noun}\}^{7}, \{,^{comma}\}^{8}, \{the^{definite\text{-}article}\}^{9}, \{lion^{noun}\}^{10}, \{sleeps^{verb}\}^{11}, \{tonight^{adverb}\}^{12}\}^{lyrics}.$$

Now you have an explanation. We can, with very little effort, enable the formation of ordered n-tuples by including a subset of the natural numbers in any genesis set we choose to create. And, in the sphere of software, there is good reason to do that, because software is often required to sort data into a particular order.

More importantly, we have arrived at the point where we can confidently declare what an ordered pair is.

It's an ordered 2-tuple – a certain set of two couplets – nothing more, nothing less.

Taking Stock

This chapter provided additional information about couplets, and then climbed the tower into the algebra of relations. The following bullet points summarize its content:

- We can use the natural numbers within a Cartesian product to create ordered sets, ordered n-tuples.
- We define the functions *yin* and *yang* as follows: *yin* $(a^b) = a$ and *yang* $(a^b) = b$. These functions also can be applied to Relations(Γ).
- We defined the yin-set and the yang-set as follows: If we apply the *yin* or *yang* function to a relation R, we call the result the yin-set of R, or the yang-set of R, accordingly.
- The transpose of a relation R is as follows:

$$\overleftrightarrow{R} = \{\, b^a \in B \times A : \overleftrightarrow{b^a} \in A \times B\}.$$

- We defined yin-functional: If $a^b \in R$ and $a^c \in R$ imply that $b = c$ then R is said to be yin-functional.
- We defined yang-functional: R is yang-functional if $\overleftrightarrow{R}$ is yin-functional.
- Composition for Couplets(Γ): $a^b \circ d^c = a^c$ if, and only if, $b = d$.
- Composition is associative but not commutative.
- The diagonal of Γ is defined as $D_\Gamma = \{a^a \in \mathrm{P}(\Gamma \times \Gamma), a \in \Gamma\}$. It is the identity for composition.
- The signature of Relations(Γ) is $[\mathrm{P}(\Gamma \times \Gamma), \{[\circ, D_\Gamma]\}, \{\leftrightarrow\}]$. The set operators, union ∪, intersection ∩, complement ′ and the subset relation ⊂ are all valid for Relations(Γ).
- We define n-tuples as follows: If $A \cup B \cup C \cup \ldots \cup N \cup U$, we designate $A \times B \times C \times \ldots \times N$ to be

$$\{\{1^a, 2^b, 3^c \ldots, n^z\} \in \mathrm{P}(\{1, 2, 3 \ldots, n\} \times U) : a \in A, b \in B, c \in C, \ldots, z \in N\}.$$

Chapter 7: The Gathering of the Clans

There are nine and sixty ways of constructing tribal lays, And every-single-one-of-them-is-right!

~ Rudyard Kipling

IN THE EARLY DAYS of the relational model of data, many of its enthusiastic advocates insisted that there was a "right structure" for data. This "right structure" was called "third normal form." The "normalizing" of data to "third normal form" was a fairly simple data modeling process that aimed to minimize redundancy (i.e., data duplication) and incorrect data dependencies, so that update anomalies could not occur.

This data modeling approach was gradually eclipsed by Peter Chen's more natural Entity Relationship (ER) model, which produced a very similar outcome and gradually became dominant. Such data models were the workhorses for the OLTP systems they helped to build through the 1980s and 1990s. However, by the late 1990s, as large companies began constructing data models for the large collections of data that inhabited a data warehouse, a fully normalized data structure proved impractical. This was due to the multitude of awkwardly slow database JOIN operations it engendered.

Consequently, an alternative approach was adopted. The data to be held in the data warehouse first needed to be normalized to third normal (or some variant thereof), and then the normalized model needed to be compressed into far fewer tables to create what was called a "snowflake schema" or even a "star schema." If this process of normalizing in order to denormalize has you reciting *The Grand Old Duke of York*, you are not alone.

Certainly there are data structures that are easier for human beings to understand. Technically, such models of data are usually referred to as logical models. These models hide many of the details of data that a computer needs to know. Sadly, humans have never achieved complete agreement on exactly what the logical model should be for a given collection of information. There are debates and disagreements.

There is also what is called a physical model of data, which arranges the data in a form that a computer can know and love. In truth, this is not so much a model as a data structure that is implemented for a given collection of information. However, it is often the case that this structure derives from a software engineer's model of the way a computer can most effectively process the data.

Here there are pragmatic reasons why there may be disagreement on the best data structure. Once data enters the computer, there is no knowing for certain how it will be used. All a software engineer can do is estimate how it may be used and structure it accordingly. The optimal data structure will be the one that consumes least computer resources while meeting the response time or throughput goals of the application. Because computer hardware is continually evolving – the power of CPUs and GPUs, continually improves, memory speeds up, SSDs speed up, networks speed up and so on – the optimal data structure for any application itself changes with time.

In practice, the data structures that computers know and love are distinctly different from the data structures that the humans know and love (or disagree about). Consequently, in building any application, it is necessary to map the logical model to the optimal physical data structure so that computers and human beings can live together in relative harmony.

From a data algebraist's perspective, there is no correct data structure. Certainly there can be incorrect structures that are algebraically deficient, but that deficiency will find its origin in human error. Where that is not the case, in situations where competing logical data models are fighting it out, we are happy to proclaim that every-single-one-of-them-is-right.

As for physical data structures, the goal is optimization, and mathematicians have skin in that game – or better put, optimization is the soul of mathematics. There is very good reason to believe that optimizing physical digital data structures, no matter what their context, is a work that data algebra will dominate in the long run.

In the meantime, almost all the world's data is structured in some way or other, and, no matter how it is structured, its structure can be captured in an algebraic form, as we are gradually demonstrating in this book. Some of the data, perhaps even most of it, may be algebraically deficient, but that can be fixed – and hopefully in time it will be. The rest of it is in the right form for the application of data algebra, but it may not be in the optimal form.

Relations beget clans

A clan is a set of relations. That's how it works in Scotland and that's how it works in data algebra. If the relations are similar as they are in the example below, then the clan is similar to a table in a relational database, and a clan can certainly express such a table.

This is a clan, $\mathbb{O}$, of Oscar-winning actresses:

$$\{\{2022^{year}, \textit{Jessica Chastain}^{actress}, \textit{The Eyes of Tammy Faye}^{movie}\}, \{2023^{year}, \textit{Michelle Yeoh}^{actress}, \textit{Everything Everywhere All at Once}^{movie}\}, \{2024^{year}, \textit{Emma Stone}^{actress}, \textit{Poor Things}^{movie}\}, \{2025^{year}, \textit{Mikey Madison}^{actress}, \textit{Anora}^{movie}\}\}.$$

You may prefer to view it using the visual artifice of a table:

Table 1: Table O of Oscar-winning actresses

year	*actress*	*movie*
2022	Jessica Chastain	The Eyes of Tammy Faye
2023	Michelle Yeoh	Everything Everywhere All at Once
2024	Emma Stone	Poor Things
2025	Mikey Madison	Anora

If you prefer that, then for algebraic accuracy you need to remember that the last four rows of the table are just the *yin* sets of the relations they represent. The column headings are, believe it or not, *yang* ($\mathbb{O}$). This provides us with an excellent excuse for explaining what the *yin* and *yang* of a clan are.

The yin-set of a clan $\mathbb{C}$, *yin*($\mathbb{C}$), is the union of the yin sets of the relations in $\mathbb{C}$. We can write this:

$$yin(\mathbb{C}) = \cup_{R \in \mathbb{C}}\, yin(R).$$

In our example, *yin*($\mathbb{O}$) is:

> *{2022, Jessica Chastain, The Eyes of Tammy Faye, 2023, Michelle Yeoh, Everything Everywhere All at Once, 2024, Emma Stone, Poor Things, 2025, Mikey Madison, Anora}.*

You already know what *yang*($\mathbb{O}$) is. The general algebraic expression is:

$$yang(\mathbb{C}) = \bigcup_{R \in \mathbb{C}} yang(R).$$

Clans do not have to be conveniently structured so that they fit neatly into a table. We can, for example, create a clan by simply lumping together two of our previous examples of relations, as follows:

$$\mathbb{A} = \{\{Thor^{hammer}, Hulk^{anger}, Iron\ Man^{suit}\},$$
$$\{Bugs^{Bunny}, Daffy^{Duck}, Wile\ E.^{Coyote}\}\}.$$

It is hard to imagine a good reason to do that, but nevertheless if you do it, you have a set of relations, and that's a clan; you cannot shoehorn it into a table with a heading and rows, unless you are willing to accept a table containing empty cells like this:

Table 2: Table with empty cells

hammer	***anger***	***suit***	***Bunny***	***Duck***	***Coyote***
Thor	*Hulk*	*Iron Man*			
			Bugs	*Daffy*	*Wile E.*

We can also build a clan directly from a relation. For example, if we take the simple relation:

$$B = \{Bugs^{Bunny}, Daffy^{Duck}, Wile\ E.^{Coyote}\}$$

and partition it into two subsets, and then collect them together in a set as shown:

$$\mathbb{B} = \{\{Bugs^{Bunny}, Daffy^{Duck}\},$$
$$\{Wile\ E.^{Coyote}\}\}$$

we have created a clan from the relation. Similarly, if we take the union of those two relations, the clan is immediately demoted to being a relation. That's the joy of the algebraic tower. In some circumstances you can climb up the tower and then climb back down, as long as you don't break any of the rules.

Transposition and composition

You will be comforted to know that operations of transposition and composition lift from Relations(Γ) to Clans(Γ) with the minimum of fuss. Let's show and tell:

If $\mathbb{A} = \{\{Thor^{hammer}, Hulk^{anger}, Iron\ Man^{suit}\}, \{Bugs^{Bunny}, Daffy^{Duck}, Wile\ E.^{Coyote}\}\}$,

then

$$\overleftrightarrow{\mathbb{A}} = \{\{hammer^{Thor}, anger^{Hulk}, suit^{Iron\ Man}\}, \{Bunny^{Bugs}, Duck^{Daffy}, Coyote^{Wile\ E.}\}\}.$$

Now that we've demonstrated it, we can confidently chant that the transpose of the clan is the clan of the transpositions, or *the-flarn-of-the-clarp-is-the-clarp-of-the-flarns.*

The same applies with composition: the composition of two clans is the clan of the compositions.

Consider a very-simple-for-the-sake-of-illustration clan $\mathbb{B}$:

$$\mathbb{B} = \{\{anger^{management}, Duck^{Dynasty}\}, \{suit^{yourself}\}\}.$$

$$\begin{aligned}\mathbb{A} \circ \mathbb{B} = \{\{\text{Thor}^{hammer}, Hulk^{anger}, Iron\ Man^{suit}\}, \{Bugs^{Bunny}, Daffy^{Duck}, \\ Wile\ E.^{Coyote}\}\} \circ \{\{anger^{management}, Duck^{Dynasty}\}, \{suit^{yourself}\}\} \\ = \{\{Hulk^{management}\}, \{Iron\ Man^{yourself}\}, \{Daffy^{Dynasty}\}\}.\end{aligned}$$

Particularly take note of this point: we can "compose" a couplet a^b or a relation R with a clan simply by promoting the couplet to be a clan by wrapping it in double braces as $\{\{a^b\}\}$ or by promoting the relation to be a clan by wrapping it in single braces as $\{R\}$. Mathematically, as shorthand, because we know we can do this, we tend not to show the braces.

Yin-functional and yang-functional

Just like transposition and composition, the concepts of yin-functional and yang-functional lift naturally to clans. We can define them as follows:

> $\mathbb{C}$ is yin-functional if each of its elements, i.e., each of its relations is yin-functional.

Just to remind you of what yin-functional means, a relation R is yin-functional if $a^b \in R$ and $a^c \in R$ imply that $b = c$.

> $\mathbb{C}$ is yang-functional if $\overleftrightarrow{\mathbb{C}}$ is yin-functional.

The clans $\mathbb{A}$ and $\mathbb{B}$ above are yin-functional since in both cases, their two relations are yin-functional. The clan $\mathbb{D}$ below is an example of a clan that is neither yang-functional nor yin-functional:

$$\mathbb{D} = \{\{hammer^{Thor}, blonde^{Thor}, suit^{Iron\ Man}\}, \{Bunny^{Bugs}, Duck^{Daffy}, Duck^{soup}\}\}.$$

The "first" relation is not yin-functional because of the *Thor* couplets, and the "second" relation is not yang-functional because of the *Duck* couplets. We are obliged to wrap first and second in quotes, tedious though it may seem, because $\mathbb{D}$ is a set, and hence there is no specific order to those relations.

We are now going to introduce two more yin/yang terms as a special reward for your perseverance in getting this far through the book. You are thinking that you need two more such yin/yang terms like a hole in the head, so let us try to convince you otherwise. The new terms, by the way, are yang-regular and yin-regular.

> The clan $\mathbb{C}$ is yang-regular if for each $R \in \mathbb{C}$, R is yang-functional and yang(R) = yang($\mathbb{C}$).

Our previous clan, $\mathbb{O}$, of recent Oscar-winning actresses, is yang-regular.

$$\{\{2022^{year}, Jessica\ Chastain^{actress}, The\ Eyes\ of\ Tammy\ Faye^{movie}\},$$
$$\{2023^{year}, Michelle\ Yeoh^{actress}, Everything\ Everywhere\ All\ at\ Once^{movie}\},$$
$$\{2024^{year}, Emma\ Stone^{actress}, Poor\ Things^{movie}\},$$
$$\{2025^{year}, Mikey\ Madison^{actress}, Anora^{movie}\}\}.$$

This means that the clan can be represented as a nice regular table with the headings: *year, actress and movie.* Take any one of the four relations in $\mathbb{O}$ and its yang-set is: {*year, actress, movie*}. Each of these relations is yang-functional, and if you take the union of their yang sets you get *yang*($\mathbb{O}$), i.e:

$$yang(\mathbb{O}) = \cup_{R \in \mathbb{O}} yang(R).$$

It is clear that for every R, *yang*(R) = *yang*($\mathbb{O}$) = {*year, actress, movie*}.

As you may have guessed:

$$\mathbb{C} \text{ is yin-regular if } \overleftrightarrow{\mathbb{C}} \text{ is yang-regular.}$$

Take a look at $\overleftrightarrow{\mathbb{O}}$ below:

$$\{\{\text{year}^{2022}, \textit{actress}^{\textit{Jessica Chastain}}, \textit{movie}^{\textit{The Eyes of Tammy Faye}}\},$$
$$\{\text{year}^{2023}, \textit{actress}^{\textit{Michelle Yeoh}}, \textit{movie}^{\textit{Everything Everywhere All at Once}}\},$$
$$\{\text{year}^{2024}, \textit{actress}^{\textit{Emma Stone}}, \textit{movie}^{\textit{Poor Things}}\},$$
$$\{\text{year}^{2025}, \textit{actress}^{\textit{Mikey Madison}}, \textit{movie}^{\textit{Anora}}\}\}.$$

The yang-set of each of the relations in $\overleftrightarrow{\mathbb{O}}$ is different and hence it is not the case that for every R, $yang(R) = yang\,(\overleftrightarrow{\mathbb{O}})$.

The Tier of Clans(Γ)

No doubt you are itching to know about the ground set of the clan tier of the tower. Aside from calamine lotion, the only known cure for that itch is to tell you about it.

First of all, we can promise you that it has a lot more in it than the comparatively anorexic P(Γ x Γ). The ground set of clans is P(P(Γ x Γ)), or as we prefer to write it, $P^2(\Gamma \times \Gamma)$.

To get an idea of how large a power set of a power set is, consider the almost unpopulated set {*a*, *b*}. If we write that out, it looks like this:

$$P(\{a, b\}) = \{\emptyset, \{a\}, \{b\}, \{a, b\}\}.$$

Writing out the power set of a power set is the kind of thing you are made to do for punishment if you behave badly in mathematics class and have to stay behind after school. You can see how difficult that can be just from glancing at the power set of a power set that has just two elements. It looks like this:

$$\begin{aligned} P^2(\{a, b\}) = \{&\emptyset, \{\emptyset\}, \{\{a\}\}, \{\{b\}\}, \{\{a, b\}\}, \\ &\{\emptyset, \{a\}\}, \{\emptyset, \{b\}\}, \{\emptyset, \{a, b\}\}, \{\{a\}, \{b\}\}, \{\{a\}, \{a, b\}\}, \{\{b\}, \{a, b\}\}, \\ &\{\emptyset, \{a\}, \{b\}\}, \{\emptyset, \{a\}, \{a, b\}\}, \{\emptyset, \{b\}, \{a, b\}\}, \{\{a\}, \{b\}, \{a, b\}\}, \\ &\{\emptyset, \{a\}, \{b\}, \{a, b\}\}\}. \end{aligned}$$

The 16 sets above are listed in order of increasing cardinality: first the empty set, then sets with one element (there are four), then sets with two elements (there are six), then sets with three elements (there are four) and finally the set with four elements. The actual number of elements in this set is given by:

$$|P^2(\{a, b\})| = 2^{(2^2)} = 16.$$

Having to write out the power set of the power set of a set with three elements qualifies as a cruel and unusual punishment under international law. You really don't have to have much of a genesis set before the cardinality of the power set of the power set gets astronomic. It doesn't just increase exponentially, it increases exponentially, exponentially. And as seasoned mathematicians will tell you, you need to take the logarithm of the logarithm of the cardinality to keep it under control.

And now for something completely different:

When we climb from the algebra of relations to the algebra of clans, we bump up against a symbol problem. You may have noticed that when we ascended from couplets to relations and then from relations to clans, we took care to define what we meant by transposition (↔) and composition (∘) in each new algebra. Because the effect of each operator was similar from one algebra to another, there was no confusion created by using exactly the same symbol.

When we climbed up into the algebra of relations we happily started to use the union, ∪, and intersection, ∩, operators not just because they applied, given that Relations(Γ) was a set algebra, but also because we were not using those symbols for any other purpose.

Well, when we climb up into Clans(Γ) the union, ∪, and intersection, ∩, operators pose a minor problem for us. As we remarked way back in Chapter 3, when you climb an algebraic tower, sometimes you don't get the outcome you hoped for.

So for the moment let us consider the relatively simple power set of a power set: $P^2(\{a, b, c\})$, and the sets $A = \{\{a, b\}, \{a, c\}\}$ and $B = \{\{a\}, \{a, b, c\}, \{a, c\}\}$that are elements of $P^2(\{a, b, c\})$. The union of these two sets is:

$$\begin{aligned} A \cup B &= \{\{a, b\}, \{a, c\}\} \cup \{\{a\}, \{a, b, c\}, \{a, c\}\} \\ &= \{\{a, b\}, \{a, c\}, \{a\}, \{a, b, c\}\} \end{aligned}$$

and the intersection is:

$$\begin{aligned} A \cap B &= \{\{a, b\}, \{a, c\}\} \cap \{\{a\}, \{a, b, c\}, \{a, c\}\} \\ &= \{\{a, c\}\}. \end{aligned}$$

You may not be aware of it yet, but there is another kind of union, called a cross-union, and another kind of intersection, called a cross-intersection. We bet you are drowning in anticipation and cannot wait to find out what they are. Wait no longer.

The cross-union can be defined like this:

> *The cross union of two sets of sets is the set of all possible unions of sets from the original sets.*

That's an awful thing to have to read, but we wrote it down anyway. If we had been just focusing on clans, it would have made much easier reading. We could have written a much simpler sentence like this:

The cross-union of two clans is the set of all possible unions of relations from within the respective clans.

The official definition we wrote is blessed with generality but cursed by inscrutability. Nevertheless, it's possible to get the idea if we pay attention to the words and follow an example. We use the symbol ▼to indicate the cross union. So, remembering that in this instance, $A = \{\{a,b\}, \{a, c\}\}$ and $B = \{\{a\}, \{a, b, c\}, \{a, c\}\}$ let's evaluate:

$$A \blacktriangledown B = \{\{a, b\} \cup \{a\}, \{a, b\} \cup \{a, b, c\}, \{a, b\} \cup \{a, c\}, \{a, c\} \cup \{a\}, \{a, c\} \cup \{a, b, c\}, \{a, c\} \cup \{a, c\}\}.$$

Take a look. That really is "the set of all possible unions of sets from the original sets." And if you work it out, it gives the result:

$$A \blacktriangledown B = \{\{a, b\}, \{a, c\}, \{a, b, c\}\}.$$

If we compare the two results, clearly $\mathbf{A \cup B \neq A \blacktriangledown B}$ in this particular case. It can be the same in some instances, but it is seldom so.

Naturally, the cross-intersection, which we denote by the symbol, ▲, arrives with an equally inscrutable definition:

The cross intersection of two sets of sets is the set of all possible intersections of sets from the original sets.

So, let's evaluate:

$$A \blacktriangle B = \{\{a, b\} \cap \{a\}, \{a, b\} \cap \{a, b, c\}, \{a, b\} \cap \{a, c\}, \{a, c\} \cap \{a\}, \{a, c\} \cap \{a, b, c\}, \{a, c\} \cap \{a, c\}\}.$$

Again, that really is "the set of all possible intersections of sets from the original sets." Work it out and you get:

$$A \blacktriangle B = \{\{a\}, \{a, b\}, \{a, c\}\}.$$

Again, if we compare the two results, clearly $A \cap B \neq A \blacktriangle B$. And again, it can be the same in some instances, but it is seldom so.

We can now apply all of this to couplets, relations and clans. So let's have some similar examples to work with, clans that are elements of $P^2(\{a^x, b^y, c^z\})$. Let

$\mathbb{A} = \{\{a^x, b^y\}, \{a^x, c^z\}\}$ and $\mathbb{B} = \{\{a^x\}, \{a^x, b^y, c^z\}, \{a^x, c^z\}\}$ and let's calculate the garden variety union:

$$\begin{aligned} \mathbb{A} \cup \mathbb{B} &= \{\{a^x, b^y\}, \{a^x, c^z\}\} \cup \{\{a^x\}, \{a^x, b^y, c^z\}, \{a^x, c^z\}\} \\ &= \{\{a^x, b^y\}, \{a^x, c^z\}, \{a^x\}, \{a^x, b^y, c^z\}\}. \end{aligned}$$

And now the cross-union:

$$\begin{aligned} \mathbb{A} \blacktriangledown \mathbb{B} = \{&\{a^x, b^y\} \cup \{a^x\}, \{a^x, b^y\} \cup \{a^x, b^y, c^z\}, \{a^x, b^y\} \cup \{a^x, c^z\}, \\ &\{a^x, c^z\} \cup \{a^x\}, \{a^x, c^z\} \cup \{a^x, b^y, c^z\}, \{a^x, c^z\} \cup \{a^x, c^z\}\}. \end{aligned}$$

In the above equation we really are taking the set of all possible unions of relations from within the respective clans. This gives us:

$$\mathbb{A} \blacktriangledown \mathbb{B} = \{\{a^x, b^y\}, \{a^x, c^z\}, \{a^x, b^y, c^z\}\}.$$

If we compare the two results, clearly $\mathbb{A} \cup \mathbb{B} \neq \mathbb{A} \blacktriangledown \mathbb{B}$. The union and the cross-union are definitely different beasts of burden, and both of them will serve us well in $P^2(\Gamma \times \Gamma)$.

Here's the intersection:

$$\begin{aligned} \mathbb{A} \cap \mathbb{B} &= \{\{a^x, b^y\}, \{a^x, c^z\}\} \cap \{\{a^x\}, \{a^x, b^y, c^z\}, \{a^x, c^z\}\} \\ &= \{\{a^x, c^z\}\}. \end{aligned}$$

And here's the cross-intersection:

$$\begin{aligned} \mathbb{A} \blacktriangle \mathbb{B} &= \{\{a^x, b^y\} \cap \{a^x\}, \{a^x, b^y\} \cap \{a^x, b^y, c^z\}, \{a^x, b^y\} \cap \{a^x, c^z\}, \\ &\qquad \{a^x, c^z\} \cap \{a^x\}, \{a^x, c^z\} \cap \{a^x, b^y, c^z\}, \{a^x, c^z\} \cap \{a^x, c^z\}\} \\ &= \{\{a^x\}, \{a^x, b^y\}, \{a^x, c^z\}\}. \end{aligned}$$

And, $\mathbb{A} \cap \mathbb{B} \neq \mathbb{A} \blacktriangle \mathbb{B}$. Which you probably expected, recognizing that we deliberately chose $\mathbb{A}$ and $\mathbb{B}$ to give results that didn't match for union and intersection.

The view from the tower

In summary, we ascended the tower of data algebra and discovered to our annoyance that the state of the union (and the intersection) were not as we might have hoped. Down there in $P(\{a^x, b^y, c^z\})$, we could happily take the union of $\{a^x, b^y\}$ with $\{b^y, c^z\}$ to get $\{a^x, b^y, c^z\}$. Climb the giddy tower and, what do you know, suddenly $\{\{a^x, b^y\}\} \cup \{\{b^y, c^z\}\} = \{\{a^x, b^y\}, \{b^y, c^z\}\}$.

In Relations(Γ), when we evaluate $\{a^x, b^y\} \cup \{b^y, c^z\}$, we get $\{a^x, b^y, c^z\}$, but in that medieval corner of the universe we process one relation at a time. With clans, we want to unite or divide hundreds, maybe even thousands, of relations with one blow. And this is what our brave new operators ▼ and ▲ will do for us. Let's illustrate a couple of simple possibilities by considering the clan $\mathbb{A}$, where:

$$\begin{aligned}\mathbb{A} = \{&\{a^1, a^2, a^3, a^4, a^5\},\\ &\{b^1, a^4, b^2\},\\ &\{c^1, c^2, a^5, c^3, c^6, a^6\},\\ &\{d^1, d^2\},\\ &\{e^1, d^1\}\}.\end{aligned}$$

If we want to insert a couplet, say, $abcde^1$, into each of the relations in this clan, all we need to do is construct a clan $\mathbb{B}$, where $\mathbb{B} = \{\{abcde^1\}\}$, and calculate:

$$\begin{aligned}\mathbb{A} \blacktriangledown \mathbb{B} = \{&\{abcde^1, a^1, a^2, a^3, a^4, a^5\},\\ &\{abcde^1, b^1, a^4, b^2\},\\ &\{abcde^1, c^1, c^2, a^5, c^3, c^6, a^6\},\\ &\{abcde^1, d^1, d^2\},\\ &\{abcde^1, e^1, d^1\}\}.\end{aligned}$$

Now consider the situation where we want to know whether any of the couplets a^5, c^4, d^2 or d^3 occur as an element of a relation in clan $\mathbb{A}$, then we can construct a clan $\mathbb{C} = \{\{a^5\}, \{c^4\}, \{d^2\}, \{d^3\}\}$ and calculate $\mathbb{A} \blacktriangle \mathbb{B}$. The result will be the clan $\{\{a^5\}, \{d^2\}\}$, which demonstrates that a^5 and d^2 can be found in relations in that clan.

The cross-intersection and cross-union turn out to be extremely useful operators. You will surely meet with them again.

The variety of data structures

In simple terms, we began with *the algebra of couplets*, which was quite limited, but nevertheless introduced us to the fundamental unit of data.

We climbed up to *the algebra of relations*, which introduced an algebraic data structure that can represent a simple one-record file or a simple record in a database.

The Gathering of the Clans

Now that we have ascended to *the algebra of clans*, we have attained an algebraic data structure that can represent a multi-record file or a table in a relational database, or a list, or an XML file with repeating structures.

We are now about to turn our attention to graphs, which are another data structure that can be represented by relations and clans, but before we do so, it is worth remarking that once we have demonstrated the algebraic representation of graphs, then we have discussed every common data structure in the field of computing (records, data sets, ordered sets, lists, arrays, associative arrays and graphs).

This does not mean that we have finished our ascent of the algebraic towers – we have not. Once we have had our say about clans, we will have to briefly discuss the collection of clans that are called *hordes*.

But first, let's talk about graphs.

Data Algebra and Graphs

Mathematically, a graph is a structure that models pairwise relationships between objects. The convention is to use the term "node" for the objects and the term "edge" for the connecting lines. Graphs can be *directed* in the sense that the edges indicate a direction, illustrated visually by an arrow head, or *simple,* where the edges do not indicate direction; they can also be *labeled* in the sense that there is data (a label) attached to each edge, or they can be unlabeled.

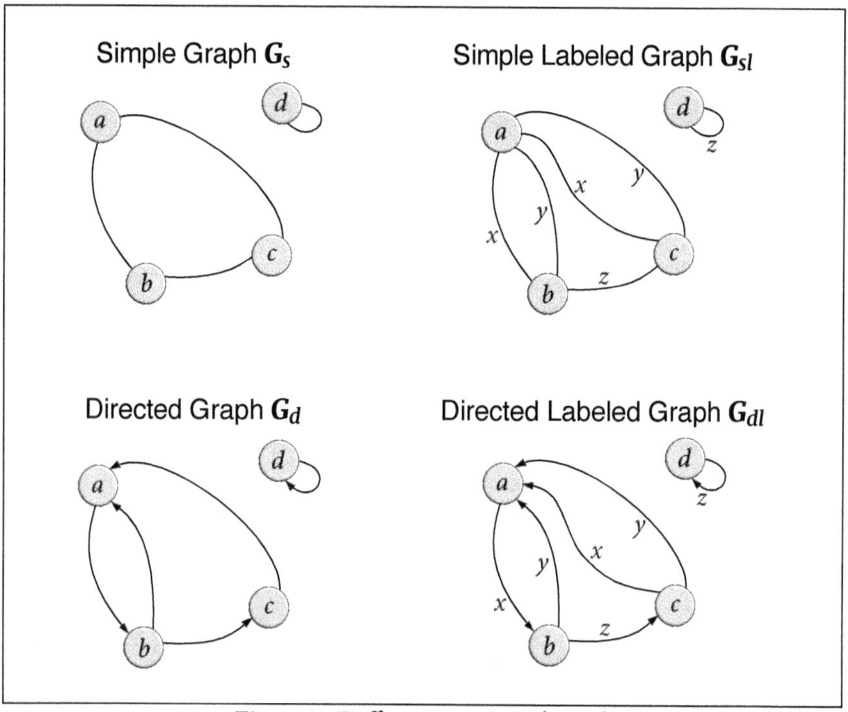

Figure 7: Different species of graph

So there are four possibilities:

1. Simple graphs
2. Simple labeled graphs
3. Directed graphs
4. Directed labeled graphs

Graphs can be represented visually, with edges connecting nodes of the graph, as illustrated in Figure 7. Such representations do not scale well, but they are useful for illustrative purposes.

There is a whole mathematics of graph theory which, if you have the time and the enthusiasm, you can investigate to your heart's content. Here we are interested primarily in graphs as data. We have a specific interest in this for two reasons:

1. Some data is better organized (for human consumption) as a graph or network of nodes.
2. Graphs expressed algebraically provide an effective way to represent semantic relationships in natural language.

Those of you who are familiar with database technology may know that relational databases are not particularly suited to graphical applications. Consequently, there are graph database products that specialize in such applications. They focus on analyzing networks of relationships between people along the lines of the popular parlor game "Six Degrees of Kevin Bacon," which is based on the relationship "who was in a movie with whom?"

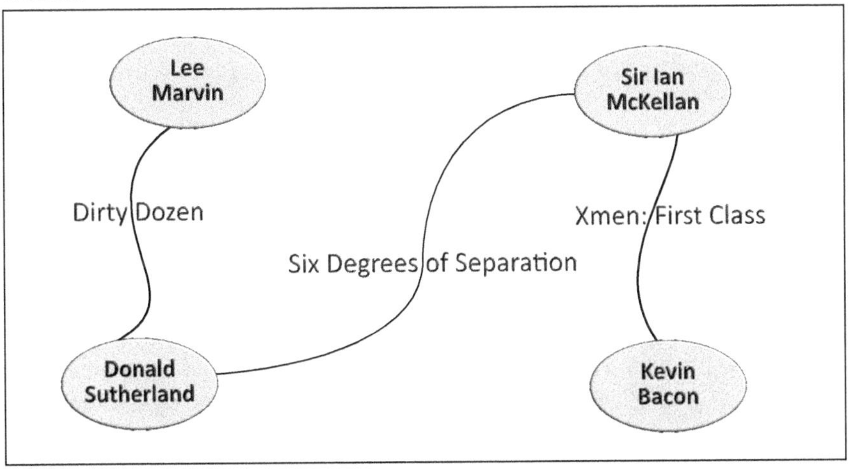

Figure 8: Six Degrees of Kevin Bacon (a simple labeled graph)

Our example in Figure 8 shows how Lee Marvin and Kevin Bacon are connected by various movies including a movie called (ironically) Six Degrees of Separation. Of course, this is simply a single answer among many possible answers drawn from a huge graph based on all the actors

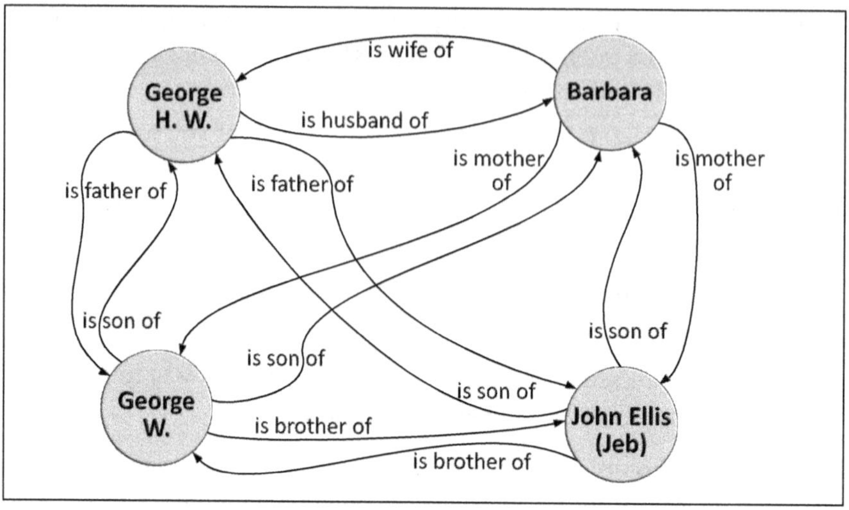

Figure 9: An example of a directed labeled graph

and actresses who ever appeared in a movie and the movies they appeared in.

The illustration in Figure 9 shows a much busier graph, with four nodes depicting the Bush family relationships between George H. W., Barbara, George W. and John Ellis (Jeb).

This graph already seems to have a few too many edges for comfort, but we could easily make it more complex by including more edges between the nodes, representing relations such as "who is older than whom," "who is taller than whom," "who is a better golfer than whom," and so on. Alternatively, we could include all of the Bush children who have never run for president.

In the first graph (Figure 8), the edges represent movies in which the two actors appeared, and the nodes represent the actors. In the second graph (Figure 9), the edges in the graph have arrowheads indicating the direction of the relationship expressed in the label and the nodes are the family members. The two data graphs are different and have different algebraic representations.

Directed labeled graphs

These graphs can have as many directed edges as you wish between nodes. We represent the adjacent directed labeled graph as:

$$\mathbf{G}_{dl} = \{(ab)x, (ba)y, (bc)z, (ca)x, (ca)y, (dd)z\}.$$

We note that:

- G_{dl} is a relation, and each of its elements is a compound couplet.
- *yang* $(a^b)^x = x$ is the label and *yin* $(a^b)^x = a^b$ is the edge.
- By convention, the source (or origin) of the edge a^b is *yin* $a^b = a$, and the destination (or target) of the edge is *yang* $a^b = b$.

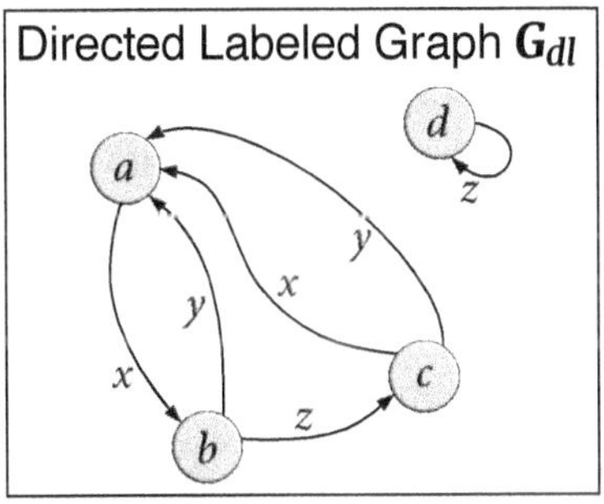

OK. We apologize in advance for the possible confusion of G's but "**G**" stands for graph and Γ (the Greek Gamma)" stands for genesis set, so please try not to confuse the two.

$\mathbf{G}_{dl}$ is a relation. If Γ is the genesis set for the elements a, b, c, x, y, etc., then we can also write:

$$\mathbf{G}_{dl} \in P((\Gamma \times \Gamma) \times \Gamma).$$

If this seems like a leap too far, remember back to when we wrote that Cartesian products create couplets. So the Cartesian product Γ x Γ can create couplets of the form a^b and thus the Cartesian product ((Γ x Γ) x Γ) can create couplets of the form $(a^b)^x$. The power set $\mathfrak{P}$((Γ x Γ) x Γ) will therefore contain all possible relations (sets of couplets) of that form.

That in turn means that P((Γ x Γ) x Γ) is a clan, because it is a set of relations. Indeed, it must be the set of all directed labeled graphs based on Γ. So we can summarize by saying that:

$$\mathbf{G}_{dl} = P((\Gamma \times \Gamma) \times \Gamma).$$

If we took our genesis set Γ to consist entirely of the Bush family tree and all possible familial relations, we would be able to create various representations of Bush Relations and various Bush clans. Although, given the terminology, we have chosen, it would probably be more fitting to involve the population of Scotland in that exercise.

Directed graphs

Now consider directed graphs with no labels. We can get to these by playing yin with G_{dl}:

$$yin\ (G_{dl}) = \{a^b, b^a, b^c, c^a, d^d\}.$$

So the yin set of a directed labeled graph is, of course, the graph without the labels:

$$G_d = \{ab, ba, bc, ca, dd\}.$$

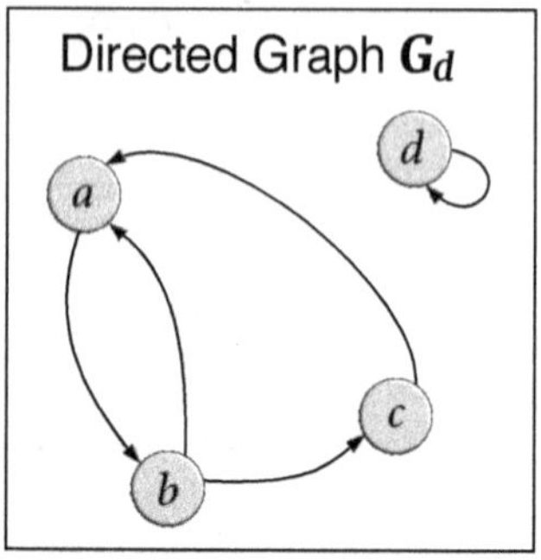

This is a simpler situation than with directed labeled graphs: $G_d \in P(\Gamma \times \Gamma)$ is just a common, garden variety relation.

An example of a graph (or network) of this kind is a map showing towns and the roads that connect them. A more interesting example, perhaps, is the map of the London Underground, which we will examine later.

Simple labeled graphs

We introduced the labeled graph with the Six Degrees of Kevin Bacon game. The movies were the labeled edges, the movie stars were the nodes and there was no direction (from star to star) involved.

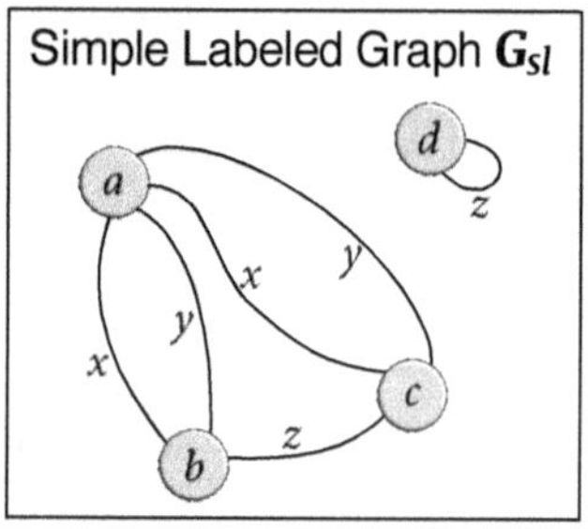

We presented our solution to the Lee Marvin–Kevin Bacon example graphically, but we could also have represented it in this way:

{{*Lee Marvin, Donald Sutherland*}$^{\textit{Dirty Dozen}}$, {*Donald Sutherland, Sir Ian McKellen*}$^{\textit{Six Degrees of Separation}}$, {*Sir Ian McKellen, Kevin Bacon*}$^{\textit{Xmen: First Class}}$}.

Our genesis set Γ for this game is the set of all movies and movie actors. The cast of any movie is a subset of Γ.

To play the Six Degrees of Kevin Bacon game, we need to pick pairs of actors, say:

{*Lee Marvin, Donald Sutherland*}$\in$P(Γ),

to serve as undirected edges. However, if we want to go beyond this specific game, say to a game that included Tony Randall and the movie *7 Faces of*

Dr. Lao, where Tony Randall plays seven different parts, we could encounter an undirected edge of the form:

$$\{\textit{Tony Randall, Tony Randall}\} = \{\textit{Tony Randall}\} \in \mathrm{P}(\Gamma),$$

indicating that in at least one movie, Tony Randall starred with Tony Randall.

Therefore, for the sake of generality, we need to identify that subset of P(Γ) consisting of all subsets of Γ of cardinality one or two:

$$\mathrm{P}(\Gamma) := \{S \in \mathrm{P}(\Gamma) : 1 \leq |S| \leq 2\}.$$

It follows that the simple labeled graph of the Six Degrees of Kevin Bacon game on page 104 and the simple labeled graph G_{sl}, indeed any simple labeled graph is an element of P(P(Γ) x Γ).

Support: another function

We derived directed graphs from directed labeled graphs using the *yin* function, so you may be wondering whether we can also derive simple labeled graphs from directed labeled graphs. If not, then pause to wonder, then continue reading.

The answer is "yes," but the mild complication is that we need to invent another function for that specific purpose. So let's do that. Observe that the natural "flow" from G_{dl} to G_{sl} is characterized by the correspondence:

$$(a^b)^x \in G_{dl} \longrightarrow \{a, b\}^x \in G_{sl}\,.$$

Notice that, in this flow, the couplet a^b uniquely determines the set $\{a, b\}$ in the following way:

$$\{a, b\} = \{yin(a^b), yang\,(a^b)\}.$$

We can capture this mathematically by referring to $\{yin\ a^b, yang\ a^b\}$ as the support of a^b and writing:

$$\boldsymbol{spt}(a^b) := \{a, b\},$$

which provides us with a useful function. But sadly, it's only a halfway house. But let us not turn our nose up at it. It's useful, as you will realize later, hopefully.

It follows, by the way, that the support of a relation is the relation of supports (*the flarn of the clarp*... etc.). For example, notice that:

$$\boldsymbol{spt}(G_d) = \{\{a, b\}, \{a, c\}, \{b, c\}, \{d, d\}\}$$
$$= \{\{a, b\}, \{a, c\}, \{b, c\}, \{d\}\}.$$

We will make use of this observation later. But let's get back to the natural flow from $\mathbf{G}_{dl}$ to $\mathbf{G}_{sl}$: $(a^b)^x \longrightarrow \{a, b\}^x$.

Clearly our *spt* function is involved here, but not "directly."

Notice in particular that:

$$\boldsymbol{spt}((a^b)^x) = \{a^b, x\}.$$

So something a little more sophisticated is called for, which is this:

$$\boldsymbol{spt}_{yin}((a^b)^x) := (\boldsymbol{spt}(a^b))^x = \{a, b\}^x.$$

It follows that spt_{yin} is a function from $(\Gamma \times \Gamma) \times \Gamma$ to $(P(\Gamma) \times \Gamma)$. This function lifts naturally to $P(\Gamma \times \Gamma) \times \Gamma$ and establishes that:

$$\boldsymbol{spt}_{yin}(G_{dl}) = G_{sl}.$$

Simple graphs

We can think of simple graphs as "labeled graph without labels." Using our examples, that is:

$$yin\ (\mathbf{G}_{sl}) = \{\{a, b\}, \{a, c\}, \{b, c\}, \{d\}\} = \mathbf{G}_s.$$

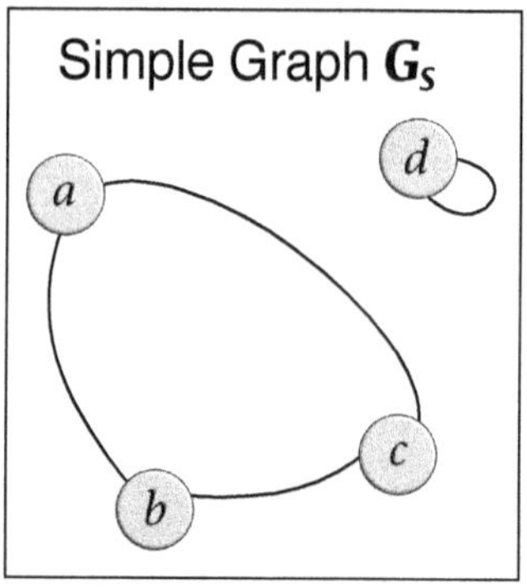

As we have just introduced the support function, it would be remiss not to draw attention to the fact that simple graphs can also be thought of as directed graphs without direction. So, we can write:

$$\boldsymbol{spt}(\mathbf{G}_l) = \{\{a, b\}, \{a, c\}, \{b, c\}, \{d\}\} = \mathbf{G}_s.$$

This is fine as far as it goes, but it leaves us with a vaguely awkward situation: $\mathbf{G}_s$ is not a relation, and because all the other graphs are, we would much prefer that it was.

How can we fix this and yet retain a mathematically faithful representation?

There is a way. Simply lift $\mathbf{G}_s$ to the diagonal of $\mathbf{G}_s \times \mathbf{G}_s$,

$$\{\{a, b\}^{\{a, b\}}, \{a, c\}^{\{a, c\}}, \{b, c\}^{\{b, c\}}, \{d\}^{\{d\}}\},$$

if the more cryptic G_S won't do.

Thus in general we can represent simple graphs as elements of:

$$P(P(\Gamma)) \text{ or } P(\boldsymbol{D}_{P(\Gamma)}),$$

$$\text{where } P(\Gamma) := \{S \in P(\Gamma) : 1 \leq |S| \leq 2\}.$$

Graphs and data

Graphs are important data structures because they are useful. A particularly good example of this is provided by maps. Maps were designed and used long before mathematicians dreamed up graph theory to systematize them. It turns out, as we hope you have gathered, that they are complex data structures.

For illustration, take a leisurely look at the example in Figure 10 on the next page – the map of The London Underground (or "the Tube," as Londoners prefer to call it). This map, of which we only show a portion, is generally regarded as a brilliant example of map design. The backstory is that Harry Beck, an engineering draftsman who worked for the London Underground, had a smoldering dislike for the geographically correct map that London Underground displayed at all its stations.

He decided to create an alternative and in 1933 came up with a map of the form illustrated. A limited number of copies of his design were printed, and it became an instant success with the public and soon displaced the previous geographically correct graph.

The London Underground was then, and still is, a complex interwoven system of different individual tube lines. Currently, there are 12 different Tube lines: The Bakerloo Line, The Central Line, The Circle Line, The District Line, The East London Line, The Hammersmith & City Line, The Jubilee Line, The Metropolitan Line, The Northern Line, The Piccadilly Line, The Victoria Line and The Waterloo & City Line. Each of these lines intersects with other lines at various stations, and they also intersect with the Docklands Light Railway, the new Queen Elizabeth line and all the main railway stations that bring commuters into London.

Beck realized that in order to navigate the Tube, the traveler's most important priority was what lines stations were on and how these lines were connected. He reasoned that actual geography didn't matter much and rather than be a slave to it, he placed the stations roughly evenly along

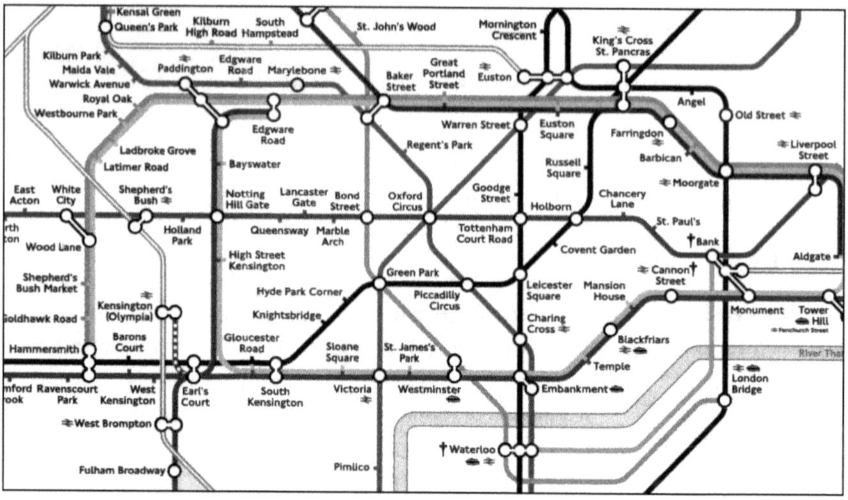

Figure 10: The map of the London Underground

lines which he ran either horizontally, vertically or at 45 degrees. He probably never thought about it in those terms, but it was an exercise in topology.

The outcome was that he shrank the map to fit it into a much smaller space with a design that was much easier to read. By squeezing and stretching the tube lines and even the River Thames itself, Beck produced a design that was eventually copied by many transport systems around the world.

Not only did this design stand the test of time, but as it evolved, more and more detail was added. Using various symbols and graphical nuances, tube stations on the map now show interchange points, disabled access, connection to river transport, connection to rail transport, airport connections and those stations closed on Sundays. The different tube lines (the edges of this graph) are color-coded, although as we are limited to a black and white illustration, this is only apparent by the different shades of gray used for the tube lines.

As I'm sure you have worked out, it qualifies as a simple labeled graph. If we take just a small central portion of the graph that includes Bond Street, Oxford Circus, Green Park and Piccadilly Circus, we can represent it with the following relation:

$\{\{Bond\ Street,\ Oxford\ Circus\}^{Central}, \{Bond\ Street,\ Green\ Park\}^{Jubilee}, \{Green\ Park,\ Oxford\ Circus\}^{Victoria}, \{Green\ Park,\ Piccadilly\ Circus\}^{Piccadilly}, \{Oxford\ Circus,\ Piccadilly\ Circus\}^{Bakerloo}\}$

In this relation, the yang of each couplet (i.e., the edge) is one of the tube lines: Central, Jubilee, Victoria or Bakerloo. The yin of the couplet is the set of cardinality two containing the two tube station names.

Imagine that you are planning to go from Bond Street to Piccadilly Circus. To do so you will need to make a mental map of the route which will involve you changing from one tube line to another. There are a number of ways that this can be achieved. One route would be to go from Bond Street to Oxford Circus on the Central Line, then change to the Victoria Line to go from Oxford Circus to Green Park, and then change onto the Piccadilly Line to go to Piccadilly Circus.

This route is a directed labeled graph:

$$\{(Bond\ Street^{Oxford\ Circus})^{Central}, (Oxford\ Circus^{Green\ Park})^{Victoria}, (Green\ Park^{Piccadilly\ Circus})^{Piccadilly}\},$$

as indeed are all other possible routes.

The yin of this route:

$$\{Bond\ Street^{Oxford\ Circus}, Oxford\ Circus^{Green\ Park}, Green\ Park^{Piccadilly\ Circus}\},$$

marks out a "path."

Graphs and semantics

Graphs can also be used in a very different context, to store the meaning of language as a collection of nodes and edges. This is becoming their most important area of application. All textual data in any language you please can be represented using directed graphs. Consider Figure 11, on the next page, which shows most of the lyrics to *Hey Jude* represented as a directed graph.

If, for example, you just wish to follow the final repeating refrain, "Na, Na, Na, Na, Na, Na, Na, Na, Na, Na, Na, Hey Jude," you can do so by following the graph, starting at the node Na. The set of couplets for this refrain is:

$$\{Na^{Na}, Na^{Hey\ Jude}\}.$$

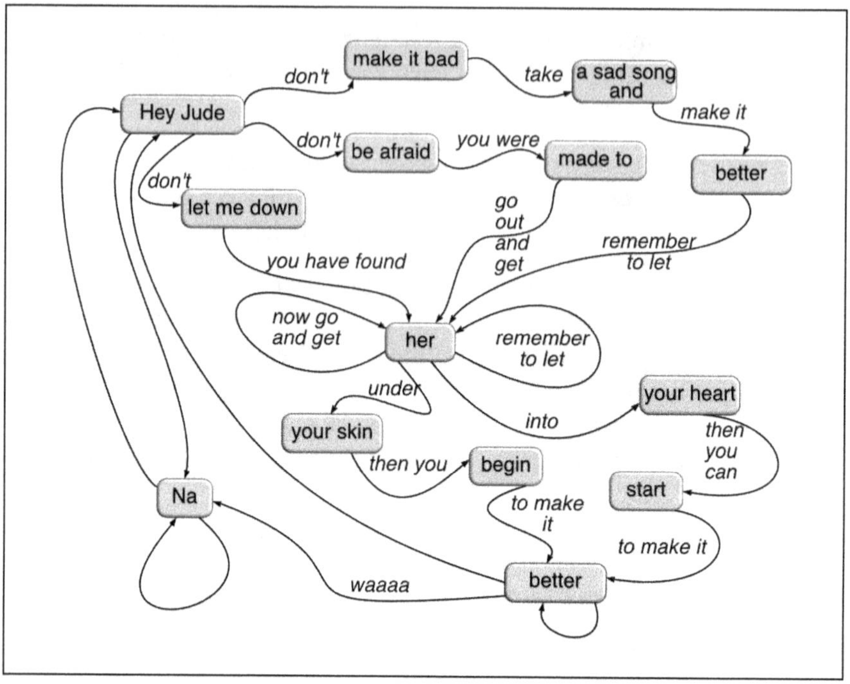

Figure 11: The lyrics to "Hey Jude," as a directed graph.

The problem with this is that it does not tell you how many times to sing "Na." So, what you might prefer is an ordered 12-tuple, like this:

$\{(Na^{Na})^{1}, (Na^{Na})^{2}, (Na^{Na})^{3}, (Na^{Na})^{4}, (Na^{Na})^{5}, (Na^{Na})^{6}, (Na^{Na})^{7}, (Na^{Na})^{8}, (Na^{Na})^{9}, (Na^{Na})^{10}, (Na^{Na})^{11}, (Na^{Hey\ Jude})^{12}\}$

to indicate the order and repetitions of "Na."

Incidentally, if the goal is to reflect the semantics of language, Figure 11 is not a good application of a directed graph. If you think in simple grammatical terms, words target other words, and words are collected together into phrases. So adjectives, prepositions and articles target nouns and adverbs target verbs. Most sentences in a language break down into triples of the form subject-predicate-object, for example:

"Data Algebra (subject) haunts (predicate) your dreams (object)."

We could write this as a triple in the following way:

$$\{Data\ Algebra^{subject}, haunts^{predicate}, your\ dreams^{object}\}.$$

If we wanted to get a little more refined and break it down into individual words, we could also represent "Data Algebra" as the couplet: $Algebra^{Data}$ and "your dreams" as the couplet $dreams^{your}$. We would then have the following triple:

$$\{(Algebra^{Data})^{subject}, haunts^{predicate}, (dreams^{your})^{object}\}.$$

The subject of a sentence doesn't have to have an object. The object can be implied as in "Data Algebra (subject) rules (predicate)." We might be tempted to write that as:

$$\{(Algebra^{Data})^{subject}, rules^{predicate}\}.$$

Having demonstrated, if only in a simple way, that we can represent textual data with directed graphs, it's probably a good time for us to introduce the topic of ontologies. If you don't know what an ontology is, shame on you! Look it up in a dictionary, and you'll be even more mystified. The dictionary will probably tell you something along the lines of:

"Ontology is the branch of metaphysics that deals with the nature of being."

This should have you wondering why the word would ever need to be plural. That definition used to be fine until Artificial Intelligence (AI) aficionados started to try to make sense of human language and needed to build semantic structures for the purpose of studying and analyzing specific "domains."

In that area of activity, an ontology is a formal, standardized explicit description of concepts (classes) in a specific domain, with properties of each concept describing associated features, attributes and constraints. Combine an ontology with a set of individual instances of classes, and you have what could be called a knowledge base. If you are not confused by that, either you're very smart or you've dozed off. Instead, think of it like this: in an area such as, say, pharmaceuticals, an ontology defines a formal semantic framework that eliminates ambiguity and enables analysis.

As you probably suspected all along, ontologies can be defined as triples.

From couplets to relations to clans (diamonds are forever)

It's time now to return to talking about graphs in a formal mathematical manner. Before we took a sight-seeing tour round the world of semantics, by way of the London Underground, we had described each of the four different kinds of graph. In each case we came to the conclusion that the graph could be represented precisely by a relation. So our different examples of graphs, G_{dl}, G_d, G_{sl} and G_s are relations.

It naturally follows that the set of all graphs of each particular kind (directed labeled, directed, simple labeled and simple) drawn from a particular genesis set Γ are clans. We can write these clans as follows:

The clan of directed labeled graphs:

$$\mathbb{G}_{dl} = P((\Gamma \times \Gamma) \times \Gamma).$$

The clan of directed graphs:

$$\mathbb{G}_{\mathbf{dl}} = P(\Gamma \times \Gamma) = \text{Relations}(\Gamma).$$

The clan of simple labeled graphs:

$$\mathbb{G}_{sl} = P(P(\Gamma) \times \Gamma),$$

$$P(\Gamma) := \{S \in P(\Gamma) : 1 \leq |S| \leq 2\}.$$

The clan of simple graphs,

$$\mathbb{G}_{s} = P(D_{P(\Gamma)}),$$

$$P(\Gamma) := \{S \in P(\Gamma) : 1 \leq |S| \leq 2\}.$$

You will have noticed, as we moved through the description of each of the four kinds of graph, that it was possible to derive the form of the elements of the less complex graphs from the elements of the more complex ones. We illustrate this in Figure 12, on the next page.

You can also think of this as mathematical calisthenics, if you like. You can manipulate these relations in the same dextrous way that you manipulated other algebraic symbols and operators in your youth.

An alternative representation

One useful point to note about graphs is that there is an alternative set-theoretical representation of them. We shall conclude this chapter by explaining it.

The relation, $G_{dl} = \{(a^b)^x, (b^a)^y, (b^c)^z, (c^a)^x, (c^a)^y, (d^d)^z\}$ can also be represented as a clan.

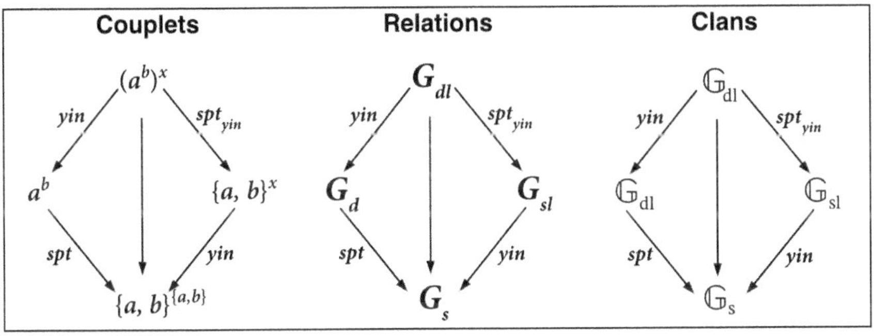

Figure 12: A display of diamonds

Here is how:

$$\begin{aligned}\mathbb{G}_{\mathrm{dl}} = \{&\{a^1, b^2, x^3\},\\ &\{b^1, a^2, y^3\},\\ &\{b^1, c^2, z^3\},\\ &\{c^1, a^2, x^3\},\\ &\{c^1, a^2, y^3\},\\ &\{d^1, d^2, z^3\}\}.\end{aligned}$$

We have taken the couplets $(a^b)^x$, etc., and represented them as ordered 3-tuples. There is a one-to-one correspondence between the couplets in the relation G_{dl} and the relations in the clan $\mathbb{G}_{\mathrm{dl}}$.

This means that we can go from one algebraic structure to the other and back, if need be. We can take some G_{dl} and transform it into a clan, $\mathbb{G}_{\mathrm{dl}}$, or take a clan $\mathbb{G}_{\mathrm{dl}}$ and transform into a relation G_{dl}. This is more mathematical calisthenics.

If you are wondering whether we can play the same trick with all the other kinds of graphs, then cease to wonder. We can. The clans that emerge are slightly different in each case, as you might imagine. And if you want to find out what they are, you'll have to work it out.

In the time-honored tradition of authors of mathematics books, we have decided to leave it as a worthy exercise for the reader.

Taking Stock

This chapter climbs the algebraic tower to consider the algebra Clans(Γ) and the lift of operators from Relations(Γ) to Clans(Γ). The following bullet points summarize its content:

- A clan is a set of relations.
- Both the operations of transposition and composition lift from Relations(Γ) to Clans(Γ) in a natural way.
- The yin-set of a clan ℂ, *yin*(ℂ), is the union of the yin sets of the relations in ℂ. We can write this:

  $$yin(\mathbb{C}) = \cup_{R \in \mathbb{C}} \; yin(R).$$

  Similarly:

  $$yang(\mathbb{C}) = \cup_{R \in \mathbb{C}} yang(R).$$

- In Clans(Γ), the union ∪ and intersection ∩ operators both apply, because it is a set algebra. However, we defined two further operators: the *cross-union* and the *cross-intersection* to offer alternative forms of union and intersection. They are defined as follows:

  The cross-union of two sets of sets is the set of all possible unions of sets from the original sets and is denoted by ▼.

  The cross-intersection of two sets of sets is the set of all possible intersections of sets from the original sets and is denoted by ▲.

- We identified four distinct types of graph: the directed labeled graph, the directed graph, the simple labeled graph and the simple graph. All of these representations of data can be defined as relations. Examples are as follows:

  Directed labeled graph: $G_{dl} = \{(a^b)^x, (b^a)^y, (b^c)^z, (c^a)^x, (c^a)^y, (d^d)^z\}$,
  Directed graph: $G_d = \{a^b, b^a, b^c, c^a, d^d\}$,
  Simple labeled graph: $G_{sl} = \{\{a, b\}^x, \{a, c\}^x, \{a, b\}^y, \{a, c\}^y, \ldots \{d\}^z\}$,
  Simple graph: $G_s = \{\{a, b\}^{\{a, b\}}, \{a, c\}^{\{a, c\}}, \{b, c\}^{\{b, c\}}, \ldots \{d\}^{\{d\}}\}$.

- Such graphs can also be represented as clans. An example is:

$$\begin{aligned}\mathbb{G}_{dl} = \{\{\, & a^1, b^2, x^3\},\\ \{\, & b^1, a^2, y^3\},\\ \{\, & b^1, c^2, z^3\},\\ \{\, & c^1, a^2, x^3\},\\ \{\, & c^1, a^2, y^3\},\\ \{\, & d^1, d^2, z^3\}\}\end{aligned}$$

Chapter 8: Joinery

It is easier to square the circle than to get round a mathematician.

~ De Morgan

SQL and the axiom of specification

You may have thought that you left all those set theory axioms behind in Chapter 3, but you didn't. One of them has returned to sit on your knee; it's the axiom of specification. This statement of the crashingly obvious proclaims that: *Every set* A *and every statement* S(**x**) *determines a unique subset* S *of* A *whose elements are exactly those elements* **x** *of* A *for which* S(**x**) *holds.*

OK, we realize the axiom's formal statement may not be aesthetically pleasing, but all it really says is that sets can have subsets, and you can get a subset by making some statement or other about elements of that set to create the subset. When it comes to real data stored in a computer, one of the ways of making such a statement is using a SQL **SELECT**.

If you are reading this book and you don't know what SQL is, we have to confess that we're surprised. But, as we are about to explore SQL, we shall allow for that possibility and explain what it is. We shall also explain every SQL statement[1] we discuss.

SQL (often pronounced "sequel") is a special-purpose language designed for accessing and managing data held in relational databases. You could call it a programming language, but it is not a general-purpose programming language like C++, Java or Python. Its capabilities are divided between data definition (defining data structures) and data manipulation (accessing and manipulating the data).

[1] *In practice, this means that the data, no matter how it is actually stored, is logically defined by a schema that represents the data as if it were stored in regular well-described tables. For convenience, in this chapter we will visually represent data in tables.*

You also need to know that there is a SQL standard, the first version of which was fathered by the American National Standards Institute (ANSI) in 1986 and adopted by the International Organization for Standardization (ISO) in 1987. Since then it has regularly been updated, and vendors of relational database products have tended to abide by it with minor variances here and there. In the SQL examples that follow, we abide by the version that Microsoft implements.[1]

The most commonly used statement in SQL is the **SELECT** statement. It enables the user or programmer to **SELECT** data from tables – in effect, taking a subset of a table of data held in the database. However, before we examine that, we'll use SQL to create a table.

The **CREATE TABLE** SQL syntax is as follows:

```
CREATE TABLE table_name
    (
    column_name1 data_type(size),
    column_name2 data_type(size),
    column_name3 data_type(size),
    ....
    );
```

The *column_name* parameters name the table columns. The *data_type* parameter specifies the column's data type (e.g., varchar, integer, date, etc.). The size parameter specifies the maximum length of the table column.

For example:

```
CREATE TABLE Outlaws
    (
    ID int,
    Name varchar(255),
    Crime varchar(255),
    Reward varchar(255),
    BornIn varchar(255),
    LastSeen varchar(255)
    );
```

If we execute this SQL statement, the outcome will be to create an empty table in a database and to update the database schema (the database's

[1] *This does not indicate a particular preference or bias; it is simply a matter of convenience as we used a Microsoft SQL reference as a source.*

formal description of the data structures it holds). The equivalent algebraic form, if we were going to read all or part of that table, would be to record the table's data structure as a relation, which we can name *OL_head*:

$$\boldsymbol{OL\_head} = \{ID^{int}, Name^{v1}, Crime^{v1}, Reward^{v1}, BornIn^{v1}, LastSeen^{v1}\}.$$

Note that above we have used the value v1 as a shorthand for the couplet varchar^{255}, where *yin*(v1) = varchar and *yang*(v1) = 255. Most databases support a variety of data types, some of which are couplets. So in general, if we are dealing with a relation *R* that defines a table row, *yang* (*R*) will be a collection of data types which may or may not itself be a relation. In this instance:

$$\boldsymbol{yang\ (OL\_head)} = \{int, varchar^{255}\},$$

which is not a relation – it is merely a set with one element that is a couplet. If you'd like a visual representation of this table header, then maybe this will satisfy you:

SQL users who are requesting data directly have little interest in the data types we show here, because their software manages all of that physical

Table 3: Outlaws table structure					
ID	***Name***	***Crime***	***Reward***	***BornIn***	***LastSeen***
Int	$varchar^{255}$	$varchar^{255}$	$varchar^{255}$	$varchar^{255}$	$varchar^{255}$

detail. They are interested in the data values that make up the rows of the table and the header elements which they understand as data descriptions.

Simple SQL select statements

If we assume that the table we created with our CREATE TABLE statement has just had some data poured into it, we might use a simple SELECT statement to retrieve some of that data. The SELECT statement has a variety of forms, the simplest of which only retrieve data from a single table. We will examine these first.

The following request asks for a number of specific columns of data to be retrieved from a table:

SELECT *column_name1, column_name2,...* FROM *table_name*;

Here's an example that queries the OUTLAWS table:

SELECT *Name, Crime, LastSeen*
FROM *Outlaws;*

The values in the table are as shown:

Take care to note that the table illustrated above and other tables shown in

Table 4: Outlaws table					
ID	***Name***	***Crime***	***Reward***	***BornIn***	***LastSeen***
1	*Robin Hood*	*Larceny*	*100 guineas*	*Locksley*	*Sherwood Forest*
2	*Billy the Kid*	*Murder*	*$500*	*New York*	*On the lam*
3	*Al Capone*	*Tax Evasion*	*None*	*New York*	*Frank Nitti's speakeasy*
4	*Loki*	*Deception*	*Thor's gratitude*	*Asgard*	*On Earth goddamit!*

the rest of this chapter are merely convenient artifices to represent clans in a similar form to the form that relational databases represent the same data. So, from that perspective, the table above is a convenient visual artifice for the clan:

$\mathbb{Outlaws}$ =

$\{\{1^{ID}, \textit{Robin Hood}^{Name}, \textit{Larceny}^{Crime}, \textit{100 guineas}^{Reward}, \textit{Locksley}^{BornIn}, \textit{Sherwood Forest}^{LastSeen}\},$
$\{2^{ID}, \textit{Billy the Kid}^{Name}, \textit{Murder}^{Crime}, \textit{\$500}^{Reward}, \textit{New York}^{BornIn}, \textit{On the lam}^{LastSeen}\},$
$\{3^{ID}, \textit{Al Capone}^{Name}, \textit{Tax Evasion}^{Crime}, \textit{None}^{Reward}, \textit{New York}^{BornIn}, \textit{Frank Nitti's speakeasy}^{LastSeen}\},$
$\{4^{ID}, \textit{Loki}^{Name}, \textit{Deception}^{Crime}, \textit{Thor's Gratitude}^{Reward}, \textit{Asgard}^{BornIn}, \textit{On Earth goddamit!}^{LastSeen}\},$
$\{5^{ID}, \textit{Chewbacca}^{Name}, \textit{Treason}^{Crime}, \textit{Bounty}^{Reward}, \textit{Kashyyyk}^{BornIn}, \textit{Tatooine}^{LastSeen}\}\}.$

We perform the algebraic equivalent of the SQL query:

SELECT *Name, Crime, LastSeen* FROM *Outlaws;*

with a composition in the following way:

$\mathbb{C} = \mathbb{Outlaws} \circ \{\{\text{Name}^{\text{Name}}, \text{Crime}^{\text{Crime}}, \text{LastSeen}^{\text{LastSeen}}\}\}$

$$= \{\{Robin\ Hood^{Name}, Larceny^{Crime}, Sherwood\ Forest^{LastSeen}\},$$
$$\{Billy\ the\ Kid^{Name}, Murder^{Crime}, On\ the\ lam^{LastSeen}\},$$
$$\{Al\ Capone^{Name}, Tax\ Evasion^{Crime}, Frank\ Nitti's\ speakeasy^{LastSeen}\},$$
$$\{Loki^{Name}, Deception^{Crime}, On\ Earth\ goddamit!^{LastSeen}\},$$
$$\{Chewbacca^{Name}, Treason^{Crime}, Tatooine^{LastSeen}\}\}.$$

Consider just the composition of the first relation in the first clan with the only relation in the second clan:

$$\{1^{ID}, Robin\ Hood^{Name}, Larceny^{Crime}, 100\ guineas^{Reward}, Locksley^{BornIn}, Sherwood\ Forest^{LastSeen}\} \circ \{Name^{Name}, Crime^{Crime}, LastSeen^{LastSeen}\}$$
$$= \{Robin\ Hood^{Name}, Larceny^{Crime}, Sherwood\ Forest^{LastSeen}\}.$$

So the result shown emerges when we apply that composition to every relation of Outlaws.

Incidentally, it is worth noting that the SQL SELECT would normally be used to retrieve data from a database, but the examples we are giving here are of querying tables held in memory as clans. The result that the SQL statement would usually furnish would just provide the values in the rows, with the metadata (the table heading) being implicit.

A simple variant of the SQL SELECT is SELECT DISTINCT. It has the following form:

SELECT DISTINCT *column_name1, column_name2,...* FROM *table_name*;

Consider the following example.

```
SELECT DISTINCT BornIn
FROM Outlaws;
```

Again, we perform the algebraic equivalent of the SQL query with a composition in the following way:

$$\text{Outlaws} \circ \{\{BornInBornIn\}\} = \{\{LocksleyBornIn\}, \{New\ York^{BornIn}\}, \{Asgard^{BornIn}\}, \{Kashyyyk^{BornIn}\}\}.$$

Two of the rows in the Outlaws table have the value “New York” under the BornIn heading. These are reduced to a single value by the composition.

Introducing Some Completely New Operators!

SQL provides a WHERE clause to impose conditions on the query. This is one example of this form of SELECT:

SELECT * FROM *table_name* WHERE *column_name1 operator value* OR *column_name2 operator value;*

The SELECT * means "select all that qualify." The logical operators AND and OR can be used and any number of different columns can be used to qualify the query.

Here is an example of this form:

SELECT * FROM *Outlaws* WHERE *Crime* = *'Murder'* OR *Crime* = *'Treason'*;

Believe it or not, to get to these answers we need to introduce a wholly new operator that looks like this: $\triangleright$, which goes by the name of *superstriction*. As you might have suspected, it has a partner operator denoted by $\triangleleft$ and calls itself *substriction*. That isn't all we need to do. But let's begin there.

For Relations($\mathcal{G}$) we define *superstriction* as follows:

$$A \triangleright B := A \text{ if } A \supset B,$$

and we define *substriction* as follows:

$$A \triangleleft B := A \text{ if } A \subset B.$$

Take note: these are partial operators. They give a result only if one set is a subset of the other. To be specific, superstriction produces a result only if the relation to the right of the operator is a subset of the one to the left, otherwise the outcome is undefined. And substriction produces a result only if the relation to the left of the operator is a subset of the one to the right, otherwise the outcome is undefined. Consequently, both of these operators are partial operators.

It will probably be easier to see what this means using an example. So consider the relations below:

$A_1 = \{yes^{yes}, no^{no}, maybe^{yes}, perhaps^{yes}, perchance^{yes}\}$.
$B_1 = \{yes^{yes}, maybe^{yes}, perhaps^{yes}, perchance^{yes}\}$.
$B_2 = \{no^{no}, perchance^{no}, I\ seriously\ doubt\ it^{maybe}\}$.
$B_3 = \{yes^{yes}, no^{no}, maybe^{yes}, perhaps^{yes}, perchance^{yes}, who\ knows^{not\ I}\}$.

Applying the superstriction:

$$A_1 \triangleright B_1 = A_1,$$

which is the case because B_1 is clearly a subset of A_1. Also:

$$A_1 \triangleright B_2$$

is undefined, because, although B_2 contains a couplet that is also found in A_1, it has couplets that are not.

$$A_1 \triangleright B_3$$

is also undefined because B_3 contains all the couplets of A_1 but also an extra couplet that disqualifies it from providing a result.

Now peruse our next set of example relations:

$A_2 = \{kernel^{yes}, seed^{yes}, root^{yes}, germ^{yes}\}.$
$B_4 = \{kernel^{yes}, seed^{yes}, root^{yes}, germ^{yes}, who\ cares^{yes}\}.$
$B_5 = \{seed^{yes}, root^{yes}, germ^{yes}\}.$
$B_6 = \{kernel^{yes}, seed^{yes}, root^{yes}, it\ don't^{fit}\}.$

The following are examples of substrictions:

$$A_2 \triangleleft B_4 = A_2,$$

because A_2 is clearly a subset of B_4.

$$A_2 \triangleleft B_5$$

is undefined, because B_5 is a subset rather than a superset of A_2.

$$A_2 \triangleleft B_6$$

is undefined, because B_6 has an element that is not found A_2.

Now that we have defined our brand new substriction and superstriction operators, we need to lift them to Clans(Γ). Yes, indeed. We created a new pair of operators lower down in the tower of data algebra just so we could lift them.

And the really cool thing about lifting them is that we don't get all those troublesome undefined results. But it isn't plain sailing.

Why? Well, as you should know by now: *the-flarn-of-the-clarp-is-very-often-the-clarp-of-the-flarns*. And that's how it works out in this case. The superstriction of Clans is the Clan of superstrictions. Similarly, the substriction of Clans is the Clan of substrictions.

And, while we're at it, we will need a couple of precise definitions to hang on these operators. So let's define the cross-superstriction:

> *The cross-superstriction of two clans is the clan of all the superstrictions between the relations in the original clans, and is denoted by* ▶.

Similarly:

> *The cross-substriction of two clans is the clan of all the substrictions between the relations in the original clans, and is denoted by* ◀.

Let us provide examples so you can see what's going on. Consider clans $\mathbb{A}$ and $\mathbb{B}$, where:

$$\begin{aligned}\mathbb{A} = \{&\{\textit{Billy the Kid}^{Name}, \textit{Murder}^{Crime}, \textit{\$500}^{Reward}\},\\ &\{\textit{Chewbacca}^{Name}, \textit{Treason}^{Crime}, \textit{Bounty}^{Reward}\},\\ &\{\textit{Jack Sparrow}^{Name}, \textit{Piracy}^{Crime}, \textit{\$1000 guineas}^{Reward}\},\\ &\{\textit{Mata Hari}^{Name}, \textit{Espionage}^{Crime}, \textit{untold secrets}^{Reward}\}\}\end{aligned}$$

and:

$$\begin{aligned}\mathbb{B} = \{&\{\textit{Jack Sparrow}^{Name}, \textit{Piracy}^{Crime}\},\\ &\{\textit{Mata Hari}^{Name}, \textit{Espionage}^{Crime}\},\\ &\{\textit{Bernie Madoff}^{Name}, \textit{Embezzlement}^{Crime}\}\}.\end{aligned}$$

The cross-superstriction of the two clans, $\mathbb{A}$ and $\mathbb{B}$ is:

$$\begin{aligned}\mathbb{A} \blacktriangleright \mathbb{B} = \{&\{\textit{Jack Sparrow}^{Name}, \textit{Piracy}^{Crime}, \textit{\$1000 guineas}^{Reward}\},\\ &\{\textit{Mata Hari}^{Name}, \textit{Espionage}^{Crime}, \textit{untold secrets}^{Reward}\}\}.\end{aligned}$$

And, the cross-substriction of the two clans $\mathbb{A}$ and $\mathbb{B}$ is:

$$\begin{aligned}\mathbb{B} \blacktriangleleft \mathbb{A} = \{&\{\textit{Jack Sparrow}^{Name}, \textit{Piracy}^{Crime}\},\\ &\{\textit{Mata Hari}^{Name}, \textit{Espionage}^{Crime}\}\}.\end{aligned}$$

If you remember rightly, a few pages back we set out trying to answer a SQL query and discovered that our beloved data algebra was short of an operator or two or three or four. Trying desperately not to admit that to you, the reader, we hastily descended the tower and created some convenient operators to help us out, and then we came back.

This is the tried and true mathematical tactic of: "take your problem some place else, do something there and then come back." We shimmied down the tower to Relations(Γ), created some new operators, put them in a bag, then clawed our way back up into Clans(Γ) where, after a little cross-pollination, we had what we wanted. It only remains for us to demonstrate that these operators are indeed what we need. So let's now take a swing at that SQL SELECT statement that caused all the trouble:

SELECT * FROM *Outlaws* WHERE *Crime* = *'Murder'* OR *Crime* = *'Treason'*;

We calculate the required clan ℂ in the following way:

$$
\begin{aligned}
\mathbb{C} &= \mathbb{Outlaws} \blacktriangleright \{\{\text{Murder}^{\text{Crime}}\}, \{\text{Treason}^{\text{Crime}}\}\} \\
&= \{\{2^{\text{ID}}, \textit{Billy the Kid}^{\textit{Name}}, \textit{Murder}^{\textit{Crime}}, \textit{\$500}^{\textit{Reward}}, \textit{New York}^{\textit{BornIn}}, \\
&\quad \textit{On the lam}^{\textit{LastSeen}}\}, \\
&\quad \{5^{\text{ID}}, \textit{Chewbacca}^{\textit{Name}}, \textit{Treason}^{\textit{Crime}}, \textit{Bounty}^{\textit{Reward}}, \textit{Kashyyyk}^{\textit{BornIn}}, \\
&\quad \textit{Tatooine}^{\textit{LastSeen}}\}\}.
\end{aligned}
$$

If you take another look at the Outlaws table (page 122), you can see that this clan provides the equivalent result to performing the above SQL SELECT on that table, furnishing the table rows for that heinous murderer Billy the Kid and the ill-tempered treasonous Chewbacca.

We can now change the subject and write a few words about SQL operators.

SQL operators

The SQL *operator* in the SQL statement:

SELECT * FROM *table_name* WHERE *column_name1 operator value* OR *column_name2 operator value;* can be any one of those shown in the adjacent table.

Table 5: SQL Operators

Operator	**Description**
=	*Equal*
<>	*Not equal*
>	*Greater than*
<	*Less than*
>=	*Greater than or equal*
<=	*Less than or equal*
BETWEEN	*Between an inclusive range*
LIKE	*Search for a pattern*
IN	*To specify a set of multiple possible values*

Note that some of the operators shown (<, >, etc.) presume that the values in a column have a natural order, which may not be the case. In practice, specific database products have specific rules for sorting data (collation rules), which they automatically apply to all SQL requests.

If we know that there is a valid ordering for a given table column, we can easily construct a corresponding algebraic equation that parallels SQL statements of this kind. An example might be as follows:

SELECT * FROM *Outlaws* WHERE *BornIn* BETWEEN *"Asgard"* AND *"Kashyyyk"*;

We can evaluate this with the following steps. First we evaluate:

$$\begin{aligned}\mathbb{C}_1 &= \mathbb{Outlaws} \circ \{\{BornIn^{BornIn}\}\} \\ &= \{\{Locksley^{BornIn}\}, \\ &\quad \{New\ York^{BornIn}\}, \\ &\quad \{Asgard^{BornIn}\}, \\ &\quad \{Kashyyyk^{BornIn}\}\}\end{aligned}$$

Then we create:

$$\begin{aligned}\mathbb{C}_2 &:= \{c^{BornIn} \in \mathbb{C}_1 : c \geq \text{"Asgard"} \text{ and } c \leq \text{"}Kashyyyk\text{"}\} \\ &= \{\{Asgard^{BornIn}\}, \{Kashyyyk^{BornIn}\}\}\end{aligned}$$

We then evaluate:

$$\begin{aligned}\mathbb{C}_3 &= \mathbb{Outlaws} \blacktriangleright \mathbb{C}_2 \\ &= \mathbb{Outlaws} \blacktriangleright \{\{Asgard^{BornIn}\}, \{Kashyyyk^{BornIn}\}\} \\ &= \{\{4^{ID}, Loki^{Name}, Deception^{Crime}, Thor's\ Gratitude^{Reward}, Asgard^{BornIn}, \\ &\quad On\ Earth\ goddamit!^{LastSeen}\}, \\ &\quad \{5^{ID}, Chewbacca^{Name}, Treason^{Crime}, Bounty^{Reward}, Kashyyyk^{BornIn}, \\ &\quad Tatooine^{LastSeen}\}\}.\end{aligned}$$

Keys

Before we move on to discussing **JOIN**s, we'd like to say a few words about keys, since in some circumstances (e.g., for a primary key) it is useful to designate table columns or combinations of columns as keys. So, to illustrate this, let's examine the Hunter table shown on the next page, comprising a set of individuals driven by personal motives to pursue particular targets.

The situation may arise where we want to create a unique key for this table. Given how few elements this table has, we can work out some possible solutions just by looking. With much larger clans, including clans that do not fit conveniently into tables, we require an algebraic approach that covers all possibilities.

Table 6: Hunter table				
ID	***Name***	***Title***	***Target***	***Motive***
a	*Pat Garrett*	*Sheriff*	*Billy the Kid*	*$500 reward*
b	*Bellerophon*	*Slayer*	*Chimera*	*Stheneboea's hand in marriage*
c	*Boba Fett*	*Bounty Hunter*	*Chewbacca*	*Bounty (not known)*
c	*Boba Fett*	*Bounty Hunter*	*Han Solo*	*Dislike*
d	*Elliot Ness*	*Chief Investigator*	*Al Capone*	*Dislike*

So consider the diagonal $D_{yang(\mathbb{H})}$ of the Cartesian product of $yang(\mathbb{H})$ with itself:

$$D_{yang(\mathbb{H})} = \{ID^{ID}, Name^{Name}, Title^{Title}, Target^{Target}, Motive^{Motive}\}.$$

Any subset K of $D_{yang(\mathbb{H})}$ where $|(\mathbb{H} \circ \{K\})| = |\mathbb{H}|$ is referred to as a *key* for $\mathbb{H}$.

In simple words, "a key is a subset of the diagonal where the result of its composition with $\mathbb{H}$ has the same cardinality as $\mathbb{H}$."

$D_{yang(\mathbb{H})}$ in this example has lots of subsets – 30, if you exclude the relation itself and the empty set Ø. Some of them, $\{Name^{Name}\}$ or $\{Motive^{Motive}\}$ for example, cannot be keys:

$$\mathbb{H} \circ \{Name^{Name}\} = \{\{Pat\ Garrett^{Name}\}, \{Bellerophon^{Name}\}, \{Boba\ Fett^{Name}\}, \{Elliot\ Ness^{Name}\}\}.$$

So:

$$|\mathbb{H} \circ \{Name^{Name}\}| = 4, \text{ while } |\mathbb{H}| = 5.$$

Similarly,:

$$|\mathbb{H} \circ \{Motive^{Motive}\}| = 4, \text{ while } |\mathbb{H}| = 5.$$

Moreover the relation $\{Name^{Name}, Motive^{Motive}\}$ is also a key, because:

$$|\mathbb{H} \circ \{Name^{Name}, Motive^{Motive}\}| = 5 = |\mathbb{H}|.$$

However, it may not be the most efficient key. That might be the relation $\{Target^{Target}\}$. If you check you'll see that:

$$|\mathbb{H} \circ \{Target^{Target}\}| = 5 = |\mathbb{H}|.$$

And this, as a relation, has the least possible cardinality for a key: one.

A *minimal key* for a clan is defined to be a key of least cardinality, and the minimal key for $\mathbb{H}$ is clearly $\{Target^{Target}\}$. A minimal key is likely to be the preferred choice for a key.

The SQL JOIN Family

The SQL SELECT statements that we have discussed so far were all directed against a single table. There are other forms of the SELECT statement that JOIN the data from the two tables together using the values from a column in each table. Tables *X* and *Y* below provide a simple example, showing the result:

Table 7: Table X		
noun	***adjective***	***enjoin***
good	*kosher*	*Join me*
mismatch	*unsuitable*	*ignore me*

Table 8: Table Y		
verb	***adverb***	***enjoin***
fit	*nicely*	*join me*
reject	*quickly*	*shun me*

Table 9: JOIN of X with Y using the "enjoin" column				
enjoin	***noun***	***adjective***	***verb***	***adverb***
join me	*good*	*kosher*	*fit*	*nicely*

The way the JOIN illustrated above works is that the row from table *X* that has the same value in the *enjoin* column as the row from table *Y*, "*join me*," is united with that row to create a new table that has just a single row. In the language of data algebra, we could represent this result as:

$$\{\text{good}^{\text{noun}}, \text{kosher}^{\text{adjective}}, \text{join me}^{\text{enjoin}}\} \cup \{\text{fit}^{\text{verb}}, \text{nicely}^{\text{adverb}}, \text{join me}^{\text{enjoin}}\}$$
$$= \{\text{good}^{\text{noun}}, \text{kosher}^{\text{adjective}}, \text{fit}^{\text{verb}}, \text{nicely}^{\text{adverb}}, \text{join me}^{\text{enjoin}}\}$$

Clearly JOINs are selective unions of a kind. There are many different varieties of JOIN, many of which, such as theta joins, are elaborations of simpler JOINs. We will not discuss those here. Our immediate goal is merely to demonstrate that we can do something equivalent to a SQL JOIN using data algebra.

In standard SQL there are seven types of JOIN:

- The INNER JOIN
- The LEFT JOIN
- A version of the LEFT JOIN that is exclusive
- The RIGHT JOIN
- A version of the RIGHT JOIN that is exclusive
- The OUTER JOIN
- A version of the OUTER JOIN that is exclusive.

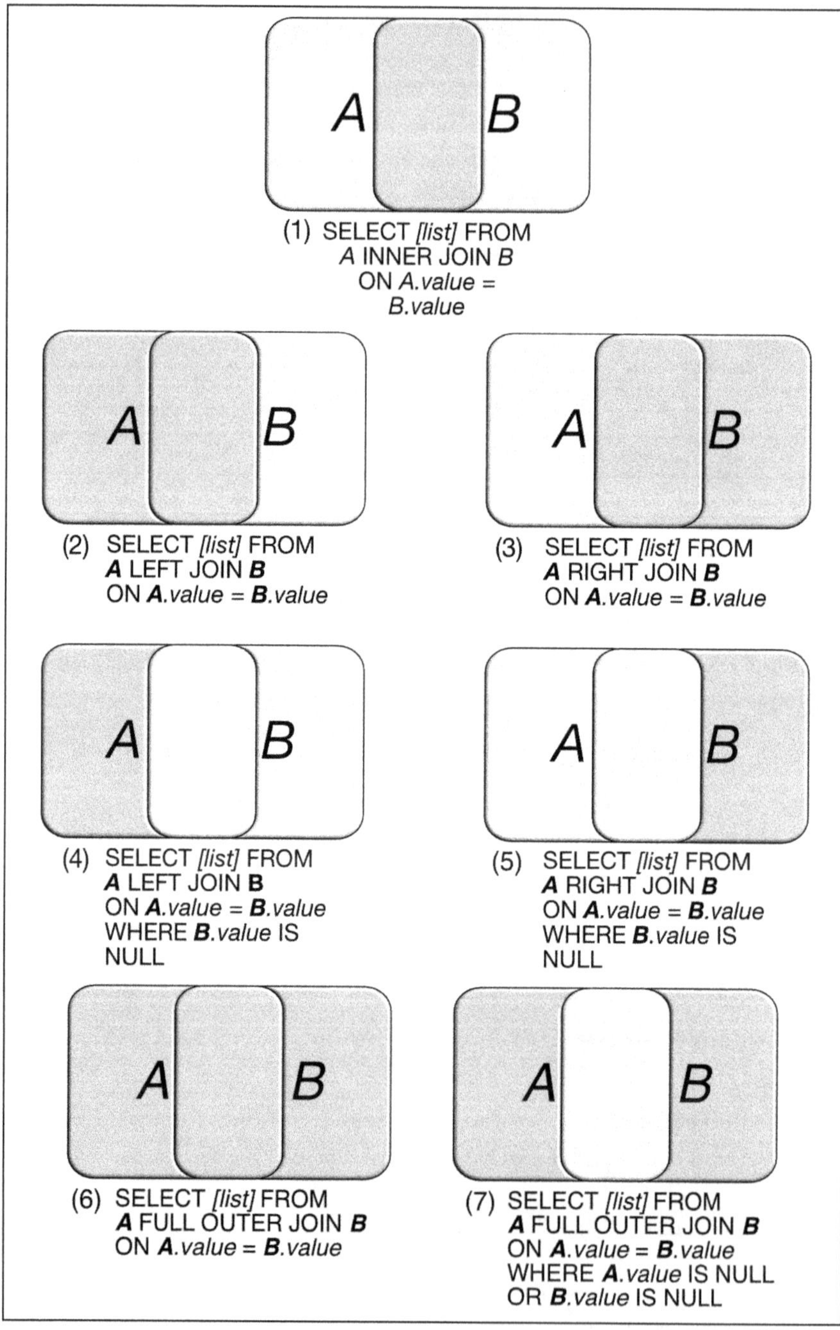

Figure 13: All possible JOINs

We illustrate all of these in Figure 13, along with the corresponding SQL SELECT statements. It has become common in SQL texts to use Venn diagrams to represent the different **JOIN**s. As this is misleading (a **JOIN** is not a corresponding set operation to any of the set theory operators that Venn diagrams represent), we've chosen to use a slightly different visual artifice, preferring rounded squares to circles.

We will use the following two small tables to illustrate the seven different JOINs.

Table 10: Example table A	
Num	***A_Words***
1	*dead*
6	*unexploded*
3	*joke*
5	*spam*
10	*Spanish Inquisition*

Table 11: Example table B	
Num	***B_Words***
9	*Ron Obvious*
8	*silly walks*
1	*parrot*
3	*warfare*
6	*Scotsman*

A SQL INNER JOIN statement takes the following general form:

SELECT *[list]* FROM *table_name1* INNER JOIN *table_name2* ON *table_name1.value = table_name2.value;*

You can also add a WHERE clause to the SELECT, but for the sake of simplicity we will not do that for the INNER JOIN. As can be seen from Figure 13, other types of SQL JOIN involve a WHERE statement.

An INNER JOIN on the tables *A* and *B* could be written as follows:

SELECT * FROM *A* INNER JOIN *B* ON *A.Num = B.Num;*

Table 12: INNER JOIN of table A and B		
Num	***A_Words***	***B_Words***
1	*dead*	*parrot*
6	*unexploded*	*Scotsman*
3	*joke*	*warfare*

The result, shown in the table above, is as you might expect. It joins data taken from rows of table *A* and *B* that have the same value for *Num*.

If we now consider this algebraically, we have two clans, $\mathbb{A}$ and $\mathbb{B}$:

$$\mathbb{A} = \{\{1^{Num}, dead^{A\_Words}\}, \{6^{Num}, unexploded^{A\_Words}\}, \{3^{Num}, joke^{A\_Words}\}, \{5^{Num}, spam^{A\_Words}\}, \{10^{Num}, Spanish\ Inquisition^{A\_Words}\}\}$$

and:

$$\mathbb{B} = \{\{9^{Num}, Ron\ Obvious^{B\_Words}\}, \{8^{Num}, silly\ walks^{B\_Words}\}, \{1^{Num}, parrot^{B\_Words}\}, \{3^{Num}, warfare^{B\_Words}\}, \{6^{Num}, Scotsman^{B\_Words}\}\}$$

To find the common values for the JOIN we proceed as follows:

$$\mathbb{I}_1 = (\mathbb{A} \circ \{Num^{Num}\}) \cap (\mathbb{B} \circ \{Num^{Num}\})$$
$$= \{\{1^{Num}\}, \{6^{Num}\}, \{3^{Num}\}, \{5^{Num}\}, \{10^{Num}\}\} + \{\{9^{Num}\}, \{8^{Num}\}, \{1^{Num}\}, \{3^{Num}\}, \{6^{Num}\}\}$$
$$= \{\{1^{Num}\}, \{6^{Num}\}, \{3^{Num}\}\}.$$

We can now apply a cross-superstriction to both $\mathbb{A}$ and $\mathbb{B}$ to identify the clans we require:

$$\mathbb{I}_2 = (\mathbb{A} \blacktriangleright \mathbb{I}_1)$$
$$= \{\{1^{Num}, dead^{A\_Words}\},\{6^{Num}, unexploded^{A\_Words}\}, \{3^{Num}, joke^{A\_Words}\}\}$$
$$\mathbb{I}_3 = (\mathbb{B} \blacktriangleright \mathbb{I}_1)$$
$$= \{\{1^{Num}, parrot^{B\_Words}\}, \{3^{Num}, warfare^{B\_Words}\}, \{6^{Num}, Scotsman^{B\_Words}\}\}.$$

Now we need to define the clan of all yang-functional relations:

$$\mathbb{F}_{yang} := \{\boldsymbol{R} \in \text{P}(\Gamma \times \Gamma): \boldsymbol{R} \text{ is yang-functional}\}.$$

If you do not immediately remember what yang-functional means in respect to a relation, here's a reminder:

A relation R is yang-functional if, and only if, a^x, $b^x \in R$ implies $a = b$.

The reason we are defining this useful clan is that we want to take a cross-union that includes only certain yang-functional relations. This obliges us to define a partial cross union of two clans. The definition is as follows:

For two clans $\mathbb{A}$ and $\mathbb{B}$,

$$\mathbb{A} \blacktriangledown_{F_{yang}} \mathbb{B} = (\mathbb{A} \blacktriangledown \mathbb{B}) \cap \mathbb{F}_{yang}.$$

Using our example to show how this works, we need to take the partial cross-union:

$$\mathbb{J} = \mathbb{I}_2 \blacktriangledown_{F_{yang}} \mathbb{I}_3 = (\mathbb{I}_2 \blacktriangledown \mathbb{I}_3) \cap \mathbb{F}_{yang}.$$

$$\begin{aligned}
\mathbb{I}_2 \blacktriangledown \mathbb{I}_3 = \{ & \{1^{Num}, dead^{A_Words}\} \cup \{1^{Num}, parrot^{B_Words}\}, \\
& \{1^{Num}, dead^{A_Words}\} \cup \{3^{Num}, warfare^{B_Words}\}, \\
& \{1^{Num}, dead^{A_Words}\} \cup \{6^{Num}, Scotsman^{B_Words}\}, \\
& \{6^{Num}, unexploded^{A_Words}\} \cup \{1^{Num}, parrot^{B_Words}\}, \\
& \{6^{Num}, unexploded^{A_Words}\} \cup \{3^{Num}, warfare^{B_Words}\}, \\
& \{6^{Num}, unexploded^{A_Words}\} \cup \{6^{Num}, Scotsman^{B_Words}\}, \\
& \{3^{Num}, joke^{A_Words}\} \cup \{1^{Num}, parrot^{B_Words}\}, \\
& \{3^{Num}, joke^{A_Words}\} \cup \{3^{Num}, warfare^{B_Words}\}, \\
& \{3^{Num}, joke^{A_Words}\} \cup \{6^{Num}, Scotsman^{B_Words}\}\} \\
= \{ & \{1^{Num}, dead^{A_Words}, parrot^{B_Words}\}, \\
& \{1^{Num}, dead^{A_Words}, 3^{Num}, warfare^{B_Words}\}, \\
& \{1^{Num}, dead^{A_Words}, 6^{Num}, Scotsman^{B_Words}\}, \\
& \{6^{Num}, unexploded^{A_Words}, 1^{Num}, parrot^{B_Words}\}, \\
& \{6^{Num}, unexploded^{A_Words}, 3^{Num}, warfare^{B_Words}\}, \\
& \{6^{Num}, unexploded^{A_Words}, Scotsman^{B_Words}\}, \\
& \{3^{Num}, joke^{A_Words}, 1^{Num}, parrot^{B_Words}\}, \\
& \{3^{Num}, joke^{A_Words}, warfare^{B_Words}\}, \\
& \{3^{Num}, joke^{A_Words}, 6^{Num}, Scotsman^{B_Words}\}\}.
\end{aligned}$$

So:

$$\begin{aligned}
\mathbb{J} &= (\mathbb{I}_2 \blacktriangledown \mathbb{I}_3) \cap \mathbb{F}_{yang} \\
&= \{\{1^{Num}, dead^{A_Words}, parrot^{B_Words}\}, \\
&\quad \{6^{Num}, unexploded^{A_Words}, Scotsman^{B_Words}\}, \\
&\quad \{3^{Num}, joke^{A_Words}, warfare^{B_Words}\}\}.
\end{aligned}$$

The intersection removes the relations in the clan $\mathbb{I}_2 \blacktriangledown \mathbb{I}_3$ that are not yang-functional. For example, $\{1^{Num}, dead^{A\_Words}, 3^{Num}, warfare^{B\_Words}\}$ has two couplets 1^{Num} and 3^{Num} that have the same yang value, but their yins, 1 and 3, are unequal.

We can now provide a general algebraic expression that gives the *natural join*, which we denote by $\bowtie$, of two clans joined by a single diagonal element:

$$\mathbb{A} \bowtie \mathbb{B} := (\mathbb{A} \blacktriangleright [\mathbb{A} \circ \{\{x^x\}\}]) \blacktriangledown_{\mathbb{F}_{yang}} (\mathbb{B} \blacktriangleright [\mathbb{B} \circ \{\{x^x\}\}])$$

$$\text{where } x \in (yang\,(\mathbb{A}) \cap yang\,(\mathbb{B})) \text{ and}$$

$$\mathbb{F}_{yang} := \{\boldsymbol{R} \in \mathrm{P}(\Gamma \times \Gamma) \text{: } \boldsymbol{R} \text{ is yang-functional}\}.$$

Incidentally, as we indicated above, it is possible to carry out a JOIN based on more than one diagonal element. This is more complex; so we leave it as an exercise for the utterly avid mathematical reader who delights in such exercises.

We cannot pretend that the above formula is identical to a natural join of two tables. It is an algebraic parallel. So, if you have two tables of data that you can validly represent as clans, then this formula will give you the clan representation of the INNER JOIN of those two tables on a single column.

Our formula compresses what we did to calculate the INNER JOIN of our example clans into a single expression. Let's walk through that again in words:

- We identified a common column for the JOIN. We did this by taking the intersection of the yang sets of $\mathbb{A}$ and $\mathbb{B}$, and selecting an element x from that set.
- The column headings of a table correspond to the yang set of the clan that represents the table. Note that there could be more than one such element in the intersection. Where that is the case, more than one INNER JOIN is possible. If there is no such element, no INNER JOIN is possible.
- Having determined x as an element of the intersection, we evaluate the compositions $\mathbb{A} \circ \{\{x^x\}\}$ and $\mathbb{B} \circ \{\{x^x\}\}$.
- We then use $\mathbb{A} \circ \{\{x^x\}\}$ and $\mathbb{B} \circ \{\{x^x\}\}$ to perform cross-superstrictions on $\mathbb{A}$ and $\mathbb{B}$ thereby removing the relations that cannot participate in the JOIN.
- Having done that, we can perform a yang-functional cross-union of the results of our cross-superstrictions, giving us the result we require.

The LEFT JOIN

Now that we have demonstrated the INNER JOIN, we can turn our attention to the other species of JOIN that SQL enables. Take a look at Figure 13 (page 132). The LEFT JOIN is (2) in the illustration.

The SQL statement is:

SELECT *[list]* FROM *A* LEFT JOIN *B*, ON *A.value* = *B.value*;

A LEFT JOIN on our example tables *A* and *B* could be written as follows:

SELECT * FROM *A* LEFT JOIN *B* ON *A.Num* = *B.Num;*

Table 13: LEFT JOIN result		
Num	*A_Words*	*B_Words*
1	*dead*	*parrot*
6	*unexploded*	*Scotsman*
3	*joke*	*warfare*
5	*spam*	*NULL*
10	*Spanish Inquisition*	*NULL*

The outcome would be as we illustrate in Table 13. We get the result of the INNER JOIN plus two additional table rows comprising of row values from table A and a "NULL" value where values from table B are inserted for the INNER JOIN. As you should be aware by now, we cannot mimic this completely because it is mathematical nonsense. The *NULL* is a duct tape and rubber-band technique used in relational databases for making something that doesn't fit into a two-dimensional table pretend that it can.

What we can do is create a clan that includes all of the valid data but excludes the *NULL*s. To do this we first calculate:

$$\begin{aligned}\mathbb{L} &= \mathbb{A} \blacktriangleright ((\mathbb{A} \circ Num^{Num}) \cap (\mathbb{B} \circ Num^{Num}))\\ &= \{\{1^{Num}, dead^{A\_Words}\},\\ &\quad \{6^{Num}, unexploded^{A\_Words}\},\\ &\quad \{3^{Num}, joke^{A\_Words}\}\}.\end{aligned}$$

This gives us all the relations from $\mathbb{A}$ that would participate in the INNER JOIN. We then take the complement of $\mathbb{L}$ in $\mathbb{A}$ as follows:

$$\begin{aligned}\mathbb{A} - \mathbb{L} = \{&\{5^{Num}, spam^{A\_Words}\},\\ &\{10^{Num}, Spanish\ Inquisition^{A\_Words}\}\}.\end{aligned}$$

We now take a union of this with the natural join:

$$(\mathbb{A} \bowtie \mathbb{B}) \cup (\mathbb{A} - \mathbb{L})$$

So:

$$\begin{aligned}(\mathbb{A} \bowtie \mathbb{B}) \cup (\mathbb{A} - \mathbb{L}) = \{&\{1^{Num}, dead^{A\_Words}, parrot^{B\_Words}\},\\ &\{6^{Num}, unexploded^{A\_Words}, Scotsman^{B\_Words}\},\\ &\{3^{Num}, joke^{A\_Words}, warfare^{B\_Words}\},\\ &\{5^{Num}, spam^{A\_Words}\},\\ &\{10^{Num}, Spanish\ Inquisition^{A\_Words}\}\}.\end{aligned}$$

Thus, the expression for evaluating a *NULL*-free algebraic approximation for the SQL LEFT JOIN is:

$$(\mathbb{A} \bowtie \mathbb{B}) \cup (\mathbb{A} - \mathbb{L}),$$

$$\text{where } \mathbb{L} = \mathbb{A} \blacktriangleright ((\mathbb{A} \circ Num^{Num}) \cap (\mathbb{B} \circ Num^{Num}))$$

The left outer join

The so-called "left outer join", diagram (4) in Figure 13 (page 132), is simply a LEFT JOIN that excludes the INNER JOIN. The SQL statement is as follows:

SELECT *[list]* FROM *A* LEFT JOIN *B*, ON *A.value* = *B.value* WHERE *B.value* IS NULL;

For our example the SQL would be:

SELECT * FROM *A* LEFT JOIN *B*, ON *A.Num* = *B.Num* WHERE *B.Num* IS NULL;

The outcome is:

Table 14: LEFT OUTER JOIN result

Num	*A_Words*	*B_Words*
5	*spam*	*NULL*
10	*Spanish Inquisition*	*NULL*

From a data algebra point of view, the equivalent result is:

$$\begin{aligned}\mathbb{A} - \mathbb{L} = \{&\{5^{Num}, spam^{A\_Words}\},\\ &\{10^{Num}, Spanish\ Inquisition^{A\_Words}\}\}.\end{aligned}$$

We have simply removed the natural join component of the LEFT JOIN.

The RIGHT JOIN

It should be clear to you, dear reader, that the RIGHT JOIN, diagram (3) in Figure 13 (page 132), involves a judicious "switching" of table *A* with table *B*. So the SQL statement is:

SELECT *[list]* FROM *A* RIGHT JOIN *B*, ON *A.value* = *B.value*;

A RIGHT JOIN on our example tables *A* and *B* could be written as follows:

SELECT * FROM *A* RIGHT JOIN *B* ON *A.Num* = *B.Num;*

The result table for the RIGHT JOIN is:

Table 15: RIGHT JOIN result		
Num	***A_Words***	***B_Words***
1	*dead*	*parrot*
6	*unexploded*	*Scotsman*
3	*joke*	*warfare*
9	*NULL*	*Ron Obvious*
8	*NULL*	*Silly Walks*

Similarly the algebraic equivalent just involves a judicious replacement of $\mathbb{A}$ with $\mathbb{B}$. The algebraic expression for evaluating a *NULL*-free algebraic approximation for the SQL RIGHT JOIN is:

$$(\mathbb{A} \bowtie \mathbb{B}) \cup (\mathbb{B} - \mathbb{R}),$$

$$\text{where } \mathbb{R} = \mathbb{B} \blacktriangleright ((\mathbb{A} \circ Num^{Num}) \cap (\mathbb{B} \circ Num^{Num})).$$

$$\begin{aligned}(\mathbb{A} \bowtie \mathbb{B}) \cup (\mathbb{B} - \mathbb{R}) = \{&\{1^{Num}, dead^{A\_Words}, parrot^{B\_Words}\},\\ &\{6^{Num}, unexploded^{A\_Words}, Scotsman^{B\_Words}\},\\ &\{3^{Num}, joke^{A\_Words}, warfare^{B\_Words}\},\\ &\{9^{Num}, Ron\ Obvious^{B\_Words}\},\\ &\{8^{Num}, Silly\ Walks^{B\_Words}\}\}.\end{aligned}$$

The right outer join

The right outer join is diagram (5) in Figure 13 (page 132), and as you should expect, also involves a switching of table *A* with table *B*. The SQL statement is:

SELECT *[list]* FROM *A* RIGHT JOIN *B*, ON *A.value* = *B.value* WHERE *A.value* IS NULL;

For our example the SQL would be:

SELECT * FROM *A* RIGHT JOIN *B*, ON *A.Num* = *B.Num* WHERE *A.Num* IS NULL;

The result table for the right outer join is:

Table 16: RIGHT OUTER JOIN result		
Num	***A_Words***	***B_Words***
9	*NULL*	*Ron Obvious*
8	*NULL*	*Silly Walks*

From a data algebra point of view, the equivalent result is:

$$\mathbb{B} - \mathbb{R} = \{\{9^{Num}, \textit{Ron Obvious}^{B\_Words}\}, \\ \{8^{Num}, \textit{Silly Walks}^{B\_Words}\}\}.$$

The FULL OUTER JOIN

The FULL OUTER JOIN, diagram (6) in Figure 13 (page 132), includes all the rows of both tables in one way or another. The SQL statement for it, using our example tables is:

SELECT * FROM *A* FULL OUTER JOIN *B*, ON *A.Num* = *B.Num*;

The result table is:

Table 17: FULL OUTER JOIN result		
Num	***A_Words***	***B_Words***
1	*dead*	*parrot*
6	*unexploded*	*Scotsman*
3	*joke*	*warfare*
5	*spam*	*NULL*
10	*Spanish Inquisition*	*NULL*
9	*NULL*	*Ron Obvious*
8	*NULL*	*Silly Walks*

The algebraic expression for evaluating a *NULL*-free algebraic approximation for the SQL FULL OUTER JOIN is:

$$(\mathbb{A} \bowtie \mathbb{B}) \cup (\mathbb{A} - \mathbb{L}) \cup (\mathbb{B} - \mathbb{R}).$$

In this instance, the resulting clan is:

$$\{\{1^{Num}, dead^{A\_Words}, parrot^{B\_Words}\},$$
$$\{6^{Num}, unexploded^{A\_Words}, Scotsman^{B\_Words}\},$$
$$\{3^{Num}, joke^{A\_Words}, warfare^{B\_Words}\},$$
$$\{5^{Num}, spam^{A\_Words}\},$$
$$\{10^{Num}, Spanish\ Inquisition^{A\_Words}\},$$
$$\{9^{Num}, Ron\ Obvious^{B\_Words}\},$$
$$\{8^{Num}, Silly\ Walks^{B\_Words}\}\}.$$

The EXCLUSIVE FULL OUTER JOIN

Finally, the exclusive full outer join, diagram (7) in Figure 13 (page 132) excludes the INNER JOIN. The SQL statement, using our example, is:

SELECT * FROM *A* FULL OUTER JOIN *B*, ON *A.Num* = *B.Num* WHERE *A.Num* IS NULL OR *B.Num* IS NULL;

The result table is:

Table 18: EXCLUSIVE FULL OUTER JOIN result		
Num	***A_Words***	***B_Words***
5	*spam*	*NULL*
10	*Spanish Inquisition*	*NULL*
9	*NULL*	*Ron Obvious*
8	*NULL*	*Silly Walks*

The algebraic expression for evaluating a *NULL*-free algebraic approximation for the SQL so-called "exclusive full outer join" is:

$(\mathbb{A} - \mathbb{L}) \cup (\mathbb{B} - \mathrm{R})$.

In our example:

$$(\mathbb{A} - \mathbb{L}) \cup (\mathbb{B} - \mathrm{R}) = \{\{5^{Num}, spam^{A\_Words}\},$$
$$\{10^{Num}, Spanish\ Inquisition^{A\_Words}\},$$
$$\{9^{Num}, Ron\ Obvious^{B\_Words}\},$$
$$\{8^{Num}, Silly\ Walks^{B\_Words}\}\}.$$

From an algebraic perspective, it is not really a JOIN at all; it's just a union.

For the sake of those readers who appreciate illustrations with their SQL JOINs, we show the full family of them in Figure 14.

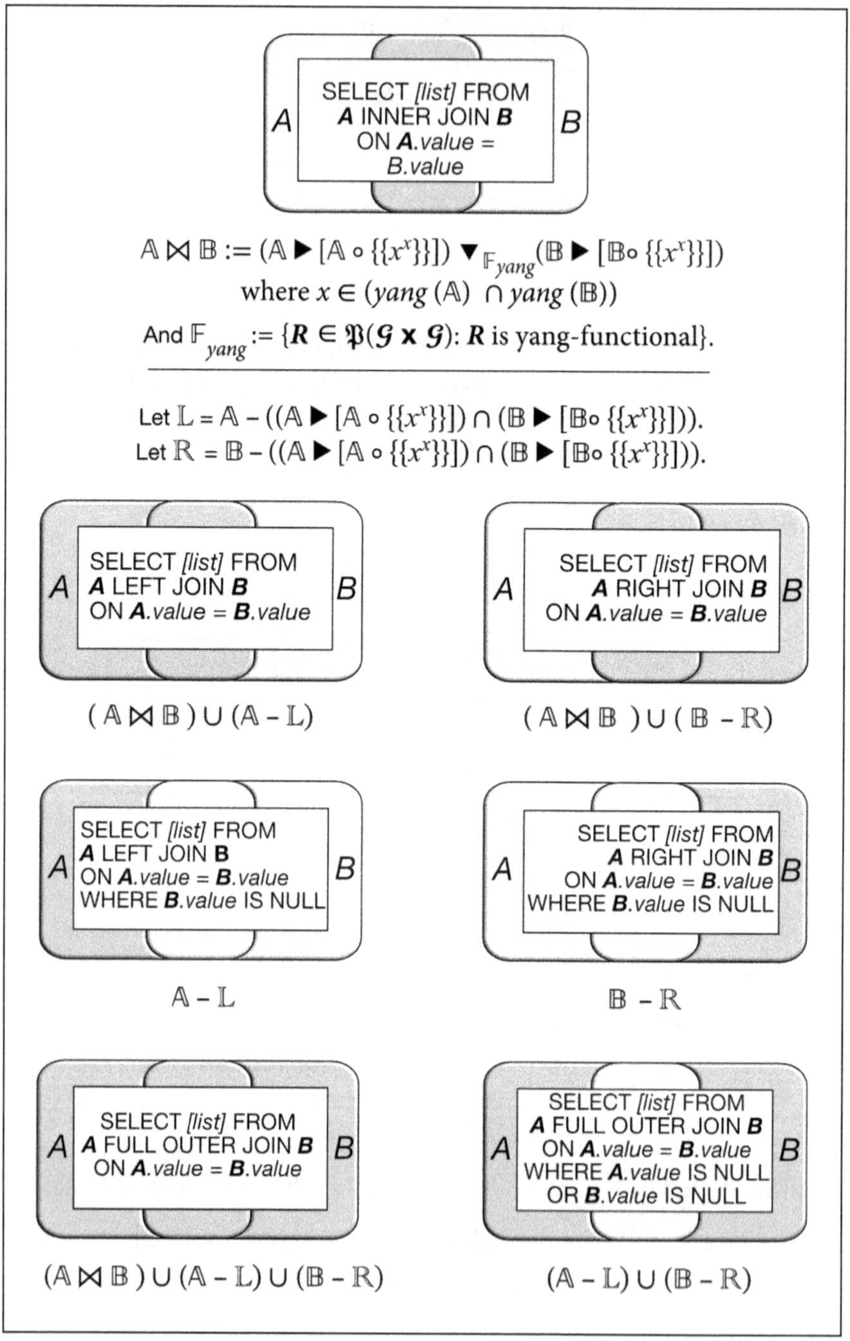

Figure 14: The family of JOINs with algebraic expressions

The Algebraic Playground

In this chapter, our intention has been to demonstrate that clans can represent tabular data structures and that such tables are subject to the rigors of clan algebra. While a clan is a mathematically legitimate more general data structure than a relational database table, such tables can be represented as clans of a particular kind. Algebraically, they are yang-regular clans. In Chapter 7, we demonstrated how graphs can be represented algebraically, both as relations and as clans.

If we consider tabular data and graphical data in all their possible forms (relational database, graphical database, RDF database, text database, document database, content databases, XML files, key value stores, flat files, etc.), it comprises a vast population of possible data structures and data collections. When we combine the three algebras: Couplets (Γ), Relations (Γ) and Clans (Γ), we have a set of mathematical tools that can represent almost all of it. Why "almost"?

Hordes

Relations are sets of couplets. Clans are sets of sets of couplets. It is possible to have sets of sets of sets of couplets. Such sets are called Hordes. Just as the ground set for relations is P(Γ x Γ) and the ground set for clans is P2(Γ x Γ), the ground set for hordes is P3(Γ x Γ), just the next tier of the tower.

Algebraically, databases are usually examples of a horde because they are sets of tables that can be represented as clans. The same can be said for a file system on almost any computer. The files may be relations or they may be clans, so the whole file system is likely to be a horde. Some XML files are hordes, as they can contain nested data structures within nested data structures. Additionally, some non-relational databases allow arrays (or tables) as elements of a data set. Data sets in such databases could be represented as hordes, or alternatively, the table could be represented as the yin of a couplet.

One of the beauties of data algebra is that each algebra is included within the algebra at the next level of the tower:

Couplets(Γ) included Relations(Γ) included Clans(Γ) included Hordes(Γ)

The level of the algebraic tower, which presides over sets of sets of sets of sets of couplets and a power set P4(Γ x Γ), is rarely visited because it is rarely needed. There are no restrictions on the elements of the genesis set Γ. The genesis set Γ is like a sliding window. There is no restriction on its size except that it cannot be an infinite set. Its elements can even include couplets or relations or clans. This allows for a extremely varied set of sophisticated data structures without even ascending to hordes.

And if any requirement emerges that necessitates the climbing of the algebraic tower up to hordes and beyond, all the operators which are used lower down at various levels of the algebraic tower – the transposition, composition, union, intersection, complementation, superstriction, substriction, cross-union, cross-intersection, cross-superstriction and cross-substriction – can be lifted to higher levels. At those higher levels, there may be other useful operators waiting to be identified.

Taking Stock

This chapter explores the use of data algebra to provide equivalent capability to the use of the SQL SELECT statements and in particular, the SQL JOIN. The following bullet points summarize its content:

A SQL SELECT that selects from a table based on column names, of the form:

- SELECT column_name1, column_name2,... FROM table_name;

is achieved by a composition of the form:

$\mathbb{C} = \mathbb{A} \circ \{\{xx, yy, ...\}\}$,

where $\mathbb{A}$ is the clan representing the table table_name and x, y, ... are values selected from the set yang($\mathbb{A}$).

In order to create parallel expressions to various SQL statements we introduced several new operators. For Relations(Γ) we defined two partial operators superstriction, denoted by ▹ and substriction, denoted by ◃, as follows:

$A \triangleright B := A \text{ if } A \supset B$

otherwise the result is undefined.

$A \triangleleft B := A \text{ if } A \subset B$

otherwise the result is undefined.

We lifted these operators to Clans(Γ) to create two further operators, the cross-superstriction, and cross-substriction. These are defined as follows:

The cross-superstriction of two clans is the clan of all the superstrictions between the relations in the original clans, and is denoted by ▸.

Also:

The cross-substriction of two clans is the clan of all the substrictions between the relations in the original clans, and is denoted by ◂.

A SQL SELECT that selects from a table based on values for particular columns, of the form:

SELECT * FROM table_name WHERE column_name1 operator value OR column_name2 operator value;

is achieved by first applying a composition then a cross-superstriction.

The general algebraic expression that gives the natural join, (equivalent to a SQL INNER JOIN) denoted by ⋈, of two clans, $\mathbb{A}$ and $\mathbb{B}$, joined by a single diagonal element is:

$(\mathbb{A} \bowtie \mathbb{B}) := (\mathbb{A} \blacktriangleright [\mathbb{A} \circ \{\{xx\}\}]) \blacktriangledown \mathbb{F}yang(\mathbb{B} \blacktriangleright [\mathbb{B} \circ \{\{xx\}\}])$

where $x \in (yang(\mathbb{A}) \cap yang(\mathbb{B}))$ and

$\mathbb{F}yang := \{R \in P(\Gamma \times \Gamma)$: R is yang-functional$\}$.

The algebraic expression for evaluating a NULL-free algebraic approximation for the SQL "left outer join" (the LEFT JOIN that excludes the INNER JOIN is:

$\mathbb{A} - ((\mathbb{A} \blacktriangleright [\mathbb{A} \circ \{\{xx\}\}]) \cap (\mathbb{B} \blacktriangleright [\mathbb{B} \circ \{\{xx\}\}]))$

- All other forms of SQL JOIN can be created by variations of the above two expressions.

Chapter 9: Unleashing The Hordes

To conquer a horde, you must first conquer yourself.

~Yoda

Another Step Up?

In the last chapter, we demonstrated in a variety of ways that clans can represent tabular data structures and that such tables are subject to the rigors of clan algebra. While a clan is a much more general data structure than a relational database table, such tables can be represented as clans of a particular kind.

Algebraically, we would say that they are yang-regular clans.

In Chapter 7, we demonstrated how graphs can be represented algebraically, both as relations and as clans. If we consider tabular data and graphical data in all their possible forms (relational database, graphical database, RDF database, text database, document database, content databases, XML files, key value stores, flat files, etc.), it comprises a vast population of possible data structures and data collections.

When we combine the three algebras: Couplets (Γ), Relations (Γ) and Clans (Γ), we have a set of mathematical tools that can represent all such data, which is almost all data.

Why "almost"?

Well, obviously we can have sets of clans. And not only can we have such sets, in computer environments we commonly have such sets. We call some of them databases. So we can have more complex sets beyond couplets, relations and clans, and as you should by now be aware, we can create an algebra for it. We call sets of clans, "Hordes".

Hordes

So the next step up the algebraic tower from Clans(Γ) is to Hordes(Γ) and a horde is a set of clans. As a clan is a set of sets of couplets, a horde must inevitably be a set of sets of sets of couplets. Any database that contains more than one clan is thus a horde, and if you think about it, most file systems are similarly hordes, as files tend to be either clans or relations.

The various operators we have discussed in this book so far could be lifted into the algebra of Hordes(Γ). But is it worth our while fully scaling the tower to identify and explain the family of operators that can populate the world of hordes?

You will be relieved to know that we are not sure about that.

Here's why there might be some reason to unleash the hordes. It is estimated that the average business desktop (a Mac or a PC) presides over somewhere between 500,000–1,500,000 files.

Of course, the operating system (OS) itself accounts for a good chunk of those: somewhere in the region of 200,000–400,000 files. Accompanying these we find enterprise software stacks (Microsoft Office, Slack, Teams, browser caches, antivirus signatures, managed endpoint agents). A typical corporate user with a few years of accumulated documents, email caches, and application data will likely have 750,000–1M files. And, of course, with developers, the numbers can easily exceed 2M.

The situation with personal desktops is not so dramatic, more in the region of 500,000-600,000 files, although extensive photo and music libraries can push that range higher.

It certainly seems like there are hordes of files. And when we examine the world of corporate servers, there can be little doubt. Corporate servers vary enormously by role, but the rough ranges are:

File servers / NAS (departmental): 5–50 million files. These accumulate relentlessly, and commonly hold 10M+ files, much of it ROT (redundant, obsolete, trivial). Studies have consistently found that 50–70% of files on corporate file shares are stale (untouched for 3+ years).

Email servers (Exchange or similar): 10–100 million objects. A 1,000-user Exchange deployment with 5 years of retention easily holds 50M+ messages and attachments, each treated as a discrete storage object.

Database servers: They hold relatively few OS-level files (thousands to low millions) but the data inside the database can represent billions of logical records. A large Oracle or SQL Server instance might have a handful of enormous datafiles on disk.

Web/application servers: 100,000–5 million files. Application code, config, logs, cached assets. Log accumulation is the main driver — an unmanaged web server can generate millions of log entries per month.

Build/CI server: 5–50 million files. Build artifacts, dependency caches, and test outputs accumulate fast. A busy Jenkins or GitHub Actions runner with poor retention policies can hit 100M+.

Backup server: 100 million–1 billion+ objects. Deduplicated backup pools (Veeam, Commvault, etc.) store enormous numbers of block-level or file-level references.

So if we consider a large enterprise and take an aggregate across all servers there are likely to be billions of files. Gartner and IDC estimate for mid-to-large enterprises the range is 1–10 billion files across all infrastructure.

We could say then, that the prosecution rests its case. There are almost uncountable hordes of files out there, your honor, and we surely need to punish them severely with their own algebra.

But wait a minute…

Maybe, with an algebra of hordes we might be biting off more than we can chew. The algebras of Couplets(Γ), Relations(Γ) and Clans(Γ) fulfill the majority of data definition and data manipulation requirements we need and it is fairly clear that these are useful. After all they allow us to define almost all the data that we ordinarily process. But anyway let us at least go to ground.

The Ground Set of Hordes

Just as the ground set for relations is $P(\Gamma \times \Gamma)$ and the ground set for clans is $P^2(\Gamma \times \Gamma)$, the ground set for hordes is $P^3(\Gamma \times \Gamma)$ – just the next tier of the tower.

As already noted, databases are usually examples of a horde and so are file systems. Some XML files are hordes, as they can contain nested data structures within nested data structures. Additionally, some non-relational databases allow arrays (or tables) as elements of a data set. Data sets in such databases are hordes, but we do not need horde algebra to manipulate such data; the table, for example, could be represented as the yin of a couplet.

And the level of the algebraic tower, which presides over sets of sets of sets of sets of couplets and a power set $P^4(\Gamma \times \Gamma)$, which we have not even honored with a label, is rarely likely to enter our thoughts.

And, of course, we can go further than that if it so pleases us, to $P^5(\Gamma \times \Gamma)$ and $P^6(\Gamma \times \Gamma)$. However, note that there are no restrictions on the elements of the genesis set Γ. The genesis set Γ is like a sliding window, with no restriction on its size except that it cannot be an infinite set. Its elements can even include couplets or relations or clans. And this allows for an extremely varied set of sophisticated data structures without even ascending to hordes.

And if any requirement emerges that necessitates the climbing of the algebraic tower to hordes and beyond, all the operators which are used lower down at various levels of the algebraic tower - the transposition, composition, union, intersection, complementation, superstriction, substriction, cross-union, cross-intersection, cross-superstriction and cross-substriction - can be lifted to higher levels. And who knows, at those higher levels, there may be other useful operators waiting to be identified.

The global datasphere stood at about 149 zettabytes as of 2024, with projections of 181 zettabytes by the end of 2025. And if you didn't know what a zettabyte is, well, it's approximately 1,024 exabytes, or using different incomprehensible numbers, it's about 1.1 trillion gigabytes. Current estimates put us on track for 230–240 zettabytes in 2026.

The global datasphere is a horde of hordes of hordes. There are vast numbers of everything, servers, web sites, applications, files and databases. Every cloud provider manages millions or gazillions of them. The algebraic tower, in principle, can represent all of it. There is no ceiling to the tower, only a diminishing need to climb it.

And by the way, there is also a natural operation that creates hordes from clans. Didn't we mention that?

If you take a clan—say, a table of sales transactions—and partition it by the values of one of its columns (perhaps by region or by quarter), the result is a horde: a set of clans, each containing the subset of transactions belonging to one partition. This is an operation that database systems perform routinely, and it is satisfying to note that the algebraic formalism captures it precisely.

Similarly, if we consider a time-versioned database — one that preserves historical snapshots of its data rather than overwriting previous states — then each snapshot is a clan and the complete history is a horde. The horde thus becomes the natural algebraic container for temporal data.

We'll have more to say about different types of database later.

Taking Stock (for the last time)

This chapter completes our giddy ascent of the algebraic tower. The following bullet points summarize its content:

- A horde is a set of clans, occupying the ground set $P^3(\Gamma \times \Gamma)$. Databases, file systems, and many other common data collections are algebraically representable as hordes.
- The algebraic tower,

  $$\text{Couplets}(\Gamma) \subset \text{Relations}(\Gamma) \subset \text{Clans}(\Gamma) \subset \text{Hordes}(\Gamma),$$

  is an inclusion chain: each level subsumes the one beneath it.
- All operators defined at lower levels of the tower (transposition, composition, union, intersection, superstriction, substriction, and their cross- variants) can be lifted to the horde level and beyond.
- Partition operations on clans naturally produce hordes. Time-versioned databases are naturally representable as hordes.
- The genesis set Γ has no restriction on its contents other than finiteness. Its elements may themselves be couplets, relations, or clans, providing enormous flexibility of representation without needing to ascend the tower.
- Higher levels of the tower, $P^4(\Gamma \times \Gamma)$ and beyond, exist but are rarely needed. The first four levels, couplets, relations, clans, and hordes, have so far proved sufficient for virtually all practical data structures.

With the mathematical foundation now complete, we turn our attention to what data algebra may be able to do in the world.

Chapter 10: Modest Math Built Our World

"Mathematics, rightly viewed, possesses not only truth, but supreme beauty."

~ Bertrand Russell

The Transformation of Data

Computers take data and transform it into other data. That is what they do, they do it all the time, and it is all that they do.

It may not feel like that is the fundamental activity that takes place when you send an email to your mother or buy an airline ticket on the web. But it is. The computers transforming the data may be cell phones, tablets, PCs or even powerful servers hiding in vast air-conditioned data centers halfway around the world. They may be communicating furiously and accessing swaths of data. But no matter what the app is and how it works, what it is doing is transforming data. That is all.

As you read these words, billions of computers are managing and processing data measured in zettabytes, and the whole digital ensemble is growing like bamboo in springtime. This voracious electronic ecosystem has swallowed up most of the information ever recorded, collected it, transformed it from one format to another and managed it. And it has spawned massive amounts of data of its own.

In its early days, the digital behemoth processed and recorded business transactions, but fairly soon it created business transactions of its own. Next it acquired an appetite for text and then became the engine for generating a great deal more. It ate its way into graphics and images and then spawned more of the same. It usurped the world of sound, consuming what existed and generating more. And finally it gobbled up video, giving birth to even more video. It consumed data, transformed data and created data. And now, in the era of large language models and generative AI, it has begun to consume the sum total of human knowledge and produce torrents of new text, images, code, and video at a pace that would have been inconceivable even a decade ago. Not only does data accumulate, data begets data.

Data Algebra and the Digital Archipelago

The Internet re-engineered this digital world in a dramatic way. From the perspective of the average Joe and Jane, it seemed to happen almost at once. Suddenly PCs had browsers, and there were search engines and commerce sites like Amazon redefining the whole idea of retail, and a vast array of web sites to explore, from the weird to the wonderful.

Although the web grew swiftly after its birth, the average Joe and Jane had little idea of how it all came about. At first, computers were islands unto themselves; exchanging data between them called for magnetic tape. Then networking was born, but it ran at a snail's pace until the advent of the PC. Of course, early PCs were also islands unto themselves, but soon there were Ethernet wires joining them to file servers and then to data center servers and mainframes.

Even so, in those dark digital days, communications between data centers was ponderous and pricey. Well, they seem like dark days from where we are looking now, a bit like European history: the Dark Ages, then the Middle Ages, then the Renaissance and then the Enlightenment.

We could think of the appearance of the magical Internet protocol as a massive stride forward to bestow connectivity on every computer that could acquire an Internet address; digital highways sprang up and criss-crossed the globe, like new trade routes.

And, as if by natural evolution, the power of computer technology and networking technology just accelerated. In almost no time at all, any person or application could, in theory, get to any data file anywhere: the Enlightenment. The realm of mobile phones and devices emerged and was assimilated, bringing us to the present day, when large language models stepped forward into the limelight.

The digital archipelago of yore has evolved to become a vast digital continent with streams of data gushing into data lakes far and wide, extending everywhere into the so-called Internet of Things–accompanied by the ultimate harvest of artificial intelligence. It is into this brave new world that Data Algebra was born.

The Modesty of Mathematics

Data Algebra? So what?

It is possible that the denizens of the IT world, the sainted system architects and the dedicated project managers and the armies of coders, will be profoundly unimpressed when they learn that there is such a thing as data algebra. They might not recognize its importance.

It is possible that their education was not embroidered with colorful examples of the taming power of mathematics and the blessings it bestowed and continues to bestow on its fields of application. Even some seasoned mathematicians may not discern the implications. That would be possible if they had never studied the profound impact that mathematics has had on the course of history. Mathematics can be accused of modesty in that respect. It is not puffed up.

Nevertheless, even a superficial analysis of mathematical discoveries and their application attests to its remarkable contribution. Set aside the fact that the ancient world would never have known how to construct its temples and its monuments if its architects and engineers had not had a good grasp of geometry and a working knowledge of mechanics. Ignore the fact that our musical system was a product of the mathematicians of antiquity, and let us cast our minds back to a date close to the commencement of the Renaissance. Think of the invention of accounting as the starting block.

Double Entry Book Keeping

It is 1494, and a man by the name of Luca Pacioli publishes a book on algebra intended for use by the schools in Northern Italy. It's not particularly snappy title was *Summa de Arithmetica, Geometria, Proportioni et Proportionalità* (Summary of Arithmetic, Geometry, Proportions and Proportionality). In a chapter entitled *Particularis de Computis et Scripturis* (Particulars of Reckoning and Writing) he explained double-entry bookkeeping, an idea which, it seems, he had invented.

His system of accounting was picked up by the Venetian merchants of the day and quickly spread from Venice to the other commercial centers of

Europe. It became the foundation for the financial management of the mercantile economies of Europe that subsequently blossomed, and it stands as the foundation of the accounting profession and as a pillar of the modern capitalistic economy. Not bad for a mere chapter in a math book.

And, incidentally, double entry bookkeeping is an idea that we will be obliged to return to a little later.

But before we go there, let us hail the achievement of John Napier, for it was he who invented logarithms in 1614. You might not be personally impressed by logarithms–maybe you never had cause to apply them. At a simple practical level, they put a calculator (log tables or a slide rule) in the hands of everyone who needed to multiply numbers as part of their occupation, and that includes scientists and engineers. The mathematical contribution of logarithms is far more extensive than that, of course, as it finds application in many areas of applied mathematics.

Dice and Card Games

Skipping along, we might like to champion the invention of probability theory. In 1654, Blaise Pascal and Pierre de Fermat exchanged a series of letters on gambling problems. They were specifically concerned with what was called "the problem of points", how to fairly divide the stakes in an unfinished game of chance–a situation that could arise when a group of gamblers became too tired to continue and wished to stagger off to bed. The problem was explained to Pascal by Antoine Gombaud, the Chevalier de Méré, a French nobleman and gambler.

No doubt you have deduced that probability and statistics are the sole foundation of the whole of the insurance industry in all its many manifestations. And it has more manifestations than you can shake a stick at. Put simply, it is enormous. In 2024, global insurance premium income was estimated to be approximately $7.8 trillion per annum, growing at about 8.6% year-over-year. Its three main segments are: life insurance at about $3 trillion in premiums, property & casualty at $2.8 trillion in premiums, and health insurance at $2 trillion in premiums.

That's not a trivial achievement for a branch of mathematics that was provoked into existence by the drunken revelry of gamblers.

The Calculus

About the same time that Blaise Pascal and Pierre de Fermat were exchanging letters, Isaac Newton and Gottfried Wilhelm Leibniz were competing for the title of "the inventor of the calculus".

We definitely abuse the word "competing" here. Newton developed his "method of fluxions" first, beginning around 1665-1666 during his annus mirabilis at Woolsthorpe, when Cambridge was closed due to plague. But he didn't publish it for decades. Newton got there first, but Leibniz published first. He arrived at his version independently around 1675-1676. He published his differential calculus in 1684 and his integral calculus in 1686. And it is Leibniz's notation (dy/dx, and the integral sign ∫) that we still use today, because it turned out to be far more practical than Newton's dot notation.

The introduction of calculus was, if anything more ground-breaking than probability and statistics. It is used by biologists to determine the growth rate of a bacterial culture, by physicists to estimate the rate of decay of radioactive elements and by chemists in thermodynamic calculations. It is employed by engineers of every persuasion and by architects, operations research analysts and statisticians. It is employed in almost every commercial activity, from mining to medicine, from communications to construction. The modern world would not exist without it.

Thus, when mathematics is introduced into a field of activity from which it was previously absent, it can engender profound changes. The historical record demonstrates this again and again. But perhaps its most dramatic demonstration is its role in the genesis of the computer.

Alan Turing

In the early twentieth century, mathematics was shaken by a sequence of discoveries that called into question its own foundations. Georg Cantor's work on infinite sets, which we have already mentioned *en passant,* had given birth to paradoxes. Bertrand Russell had shown that naive set theory led to contradictions. In response, David Hilbert–the most influential mathematician of his era–proposed what became known as the Hilbert Programme: a grand project to place the whole of mathematics on a secure, formal footing.

Hilbert's ambition was threefold. First, he wanted to show that mathematics was *consistent*—that it could never produce a contradiction. Second, he wanted to show that it was *complete*—that every true mathematical statement could, in principle, be proved. Third, and most consequentially for our story, he wanted to show that mathematics was *decidable*—that there existed a definite procedure, a mechanical method, by which one could determine, for any given mathematical statement, whether it was true or false.

This third question was called the *Entscheidungsproblem* — the "decision problem." It asked: does there exist an algorithm that can decide the truth or falsity of any statement in formal logic?

In 1931, Kurt Gödel dealt a death blow to Hilbert's programme. His incompleteness theorems demonstrated that any sufficiently powerful formal system must contain true statements that cannot be proved within the system. In other words, mathematics could not be both complete and consistent. Suck on that, Hilbert.

However, Gödel's work did not directly address the decision problem—it left open the question of whether an effective procedure for deciding provability might exist. And it was this problem of decidability that gave birth to the modern computer.

Alan Turing was twenty-three years old and a graduate student at King's College, Cambridge, when he wrote the paper that would change the world. Published in 1936 under the title *On Computable Numbers, with an Application to the Entscheidungsproblem*, it is one of the most remarkable intellectual achievements of the twentieth century—not because it answered a mathematical question, but because, in the process of answering it, Turing invented the conceptual framework for all of computing. Forget all that propaganda about Babbage, it was Turing who invented the computer.

Turing's approach was direct. To answer whether a decision procedure existed, he first had to define precisely what a "procedure" was—what it meant for a process to be mechanical, to be carried out without insight or intuition. He needed, in other words, a mathematical model of computation itself.

What he produced was an imaginary device, now called a Turing machine. It consists of an infinitely long tape divided into cells, each of which may contain a symbol. A read/write head moves along the tape, one cell at a time, reading the current symbol, writing a new one, and moving left or right, all according to a finite table of rules. The machine's behavior is entirely determined by its current state and the symbol it reads. There is no ambiguity, no room for judgement. It is mechanism distilled to its essence.

What made this construction so powerful was not the device itself—which was deliberately primitive—but what Turing proved about it. He showed that there existed a single machine, a universal Turing machine, that could simulate the behavior of any other Turing machine, provided it was given a description of that machine's rules encoded on its tape. In other words, a single device could perform any computation that any specialised device could perform, simply by reading the right instructions.

This was the birth of the programmable computer as a concept—a machine whose function is not fixed by its hardware but determined by its software.

Having defined computation rigorously, Turing was then able to prove that the Entscheidungsproblem was unsolvable. He did this by constructing an ingenious diagonal argument, showing that there existed problems— well-defined, precisely stated mathematical questions—for which no Turing machine could produce an answer in finite time. The most famous of these is the halting problem: given a description of a Turing machine and an input, determine whether the machine will eventually halt or run forever. Turing proved that no algorithm can solve this problem in general.

The philosophical implications of this are profound.

- There are things that are true but unprovable.
- There are questions that are precise but undecidable.

And that's excellent grist for the philosopher's mill, but there is also a practical consequence. By proving what could not be computed, Turing provided a complete characterization of what could be. And thus the Turing machine became the gold standard for computability. To this day, when computer scientists say that something is "computable," they mean that a Turing machine can do it.

The Code Breaker

In 1939, Britain went to war, and Turing was recruited to Bletchley Park, the secret center of British codebreaking. There, he became the central figure in the effort to break the German Enigma cipher—work that is estimated to have shortened the war by two years and saved millions of lives. (Mathematics did that.)

At Bletchley, Turing moved from the abstract to the engineered. He designed the Bombe, an electromechanical device that could rapidly test possible Enigma settings, and contributed to the design of Colossus, the first electronic computing device, built not to break the Enigma cipher, but the even more complex Lorenz cipher. These machines were not general-purpose computers in the modern sense—they ran but one program. They were specialised codebreaking tools that demonstrated that Turing's theoretical insights could be realized in physical hardware.

In the aftermath of the war, Turing worked on the design of the ACE (Automatic Computing Engine) at the National Physical Laboratory and later at the University of Manchester, where the Manchester Mark 1—one of the first stored-program computers—was developed. His 1950 paper "Computing Machinery and Intelligence," in which he proposed what is now called the Turing test, opened the field of artificial intelligence.

John von Neumann

Turing was not the only mathematician that contributed to the genesis of the computer. The other mathematical giant was John Von Neumann. It fell to him to define how a computer should be built.

John von Neumann was a force of nature—a polymath of staggering breadth whose contributions spanned pure mathematics, quantum mechanics, game theory, economics, nuclear physics, and fluid dynamics, and that was before he turned his attention to computing.

Born in 1903 in Budapest as János Neumann, he was a prodigy of almost mythological proportions. By the age of six, he could divide eight-digit numbers in his head. He received his PhD in mathematics from the University of Budapest at twenty-two, simultaneously earning a diploma in chemical engineering from ETH Zurich. By his late twenties, he was a professor at the Institute for Advanced Study in Princeton, a colleague of Einstein, Gödel, and the other luminaries who had fled Europe.

Von Neumann encountered the problem of computing through the lens of applied mathematics. During the Second World War, he was involved in the Manhattan Project, where the calculations required for the design of the implosion lens—the device that would compress the plutonium core of the atomic bomb—were so vast and complex that they pushed the limits of what human calculators and mechanical desk machines could accomplish. It was this practical bottleneck that drew von Neumann to the emerging technology of electronic computation.

The defining problem in this area concerned artillery. In those days, artillery was the god of war, and the race was on to make his thunderbolts more accurate. Funded by the U.S. Army's Ballistic Research Laboratory, a computer called ENIAC (the Electronic Numerical Integrator and Computer) was being built at the University of Pennsylvania's Moore School of Electrical Engineering.

ENIAC could perform five thousand additions per second, trivial by today's computing speeds, but revolutionary for its time. However, ENIAC had a critical limitation: the only way to program it was by physically rewiring its patch cables and setting switches. Implementing a program could take days or weeks. Von Neumann learned of ENIAC in 1944, and became interested.

In June 1945, von Neumann circulated a document titled *First Draft of a Report on the EDVAC* (EDVAC being the Electronic Discrete Variable Automatic Computer, the planned successor to ENIAC). The report, though incomplete, laid out the architecture that would come to define virtually every computer built for the next eighty years.

The key insight was the stored-program concept: the idea that a computer's instructions should be stored in the same memory as its data, encoded in the same format, and therefore capable of being manipulated — read, written, modified — by the machine itself. This was the engineering realization of Turing's universal machine. A stored-program computer does not need to be rewired to perform a different task. It simply reads different instructions from memory.

Two Legacies

Turing and von Neumann approached computing from opposite directions. Turing began with a question in mathematical logic and, in answering it, discovered the concept of the programmable computer as a theoretical inevitability. Von Neumann began with a practical need, the demand for faster calculation, and, in meeting it, gave the theoretical concept a physical form that could be manufactured, replicated, and improved.

Their contributions are complementary and inseparable. Without Turing's abstract framework, the computer would have been merely an engineering artifact, a fast abacus, with no theoretical foundation to guide its development. Without von Neumann's architectural vision, Turing's universal machine would have remained a thought experiment, a brilliant but disembodied idea.

Together, they established the two pillars on which the entire edifice of modern computing rests: the theory of computation, which tells us what can and cannot be computed, and the architecture of the stored-program computer, which tells us how to build a machine that computes. Every laptop, smartphone, server farm, and embedded microcontroller in existence today is, in essence, a physical realization of Turing's universal machine, built according to von Neumann's blueprint.

It is the greatest mathematical achievement of the twentieth century and it began with a question about the limits of mathematical proof.

Chapter 11: Good Math, Bad Math

"Medicine makes people ill, mathematics make them sad, and theology makes them sinful."

~ Martin Luther

In 1667, Johann Joachim Becher proposed that combustible materials contained a substance called terra pinguis ("fatty earth"), which was released during burning. His student Georg Ernst Stahl refined the idea around 1703 and gave the substance its famous name: phlogiston, from the Greek phlogistón ("burned up"). The theory held that all combustible materials contained phlogiston, and that burning was the process of releasing it into the air.

Wood burned because it was rich in phlogiston; ash was what remained after the phlogiston departed. Metals, when heated, lost their phlogiston and became a powdery residue called a calx (what we now call an oxide). Smelting — the reverse process — worked because charcoal, being almost pure phlogiston, restored it to the calx, regenerating the metal.

The theory was elegant, internally consistent, and accepted by virtually every chemist in Europe for the better part of a hundred years. It explained combustion, rusting, smelting, and even respiration (breathing was thought to be the body releasing phlogiston into the air). It was also wrong.

A wrong theory can be productive—phlogiston guided useful experimental work for decades and organized a great deal of empirical knowledge, even though its central premise was false.

In various fields of human endeavor you sometimes encounter a fundamental idea, which no-one questions and which is wrong. In computing, CRUD is such an idea.

CRUD is an acronym for the four fundamental operations that can be performed on persistent data: Create, Read, Update, and Delete.

The term was coined by James Martin in his 1983 book *Managing the Database Environment*, where he used it to describe the basic lifecycle of a data record. The insight behind the acronym is that, regardless of how complex a data system becomes, whether it's a flat file, a relational database, a document store, or a REST API, every interaction with stored data ultimately reduces to one of these four actions. You bring a record into existence, you retrieve it, you modify it, or you remove it. There is nothing else.

This concept has proven remarkably durable because it maps cleanly onto almost every layer of the software stack. In SQL, the four operations correspond to INSERT, SELECT, UPDATE, and DELETE. In HTTP, they map to POST, GET, PUT/PATCH, and DELETE. In file systems, they correspond to creating, opening, writing, and deleting files. It is universal, and it is wrong.

We are not blaming James Martin for this, by the way, he simply created the CRUD acronym to codify a habitual way of processing data that had been established decades before he wrote his book.

It is difficult nowadays to comprehend the constraints of early computers. At first everything was programmed in assmebler code. That was not really a language, just a set of close-to-the-metal instructions. Storage, whether in comuter memory or on punched card or magnetic tape, was expensive. Everything was expensive.

So it may have been that in the early days of computing, somebody consulted Pacioli's *Particularis de Computis et Scripturis* and consequently realized how data should be processed, but in those days computer resources were too expensive to take that route. Now they are not.

Pacioli's brilliant double entry idea created an accounting system that could be trusted. There is no U in this process. If something needs to be corrected or updated, a double entry is made that retires the old value and repalces it with the new. In other words a version of what the value was is retained, even though a new value is applied.

An update deletes data by overwriting it with a new value, and when you do that, you lose the old data value.

The delete is also a wrong idea if you truly destroy the information record by deleting it. (Most database systems log deleted values, so the infroma-tion is not lost, but files systems in general are not so careful.

CRUD is bad math. It is wrong.

The acronym should be CARD:

Create, Add, Read, Delete,

where the delete does not delete data but instead tags the data as no longer current.

Many problems arise because of CRUD, but we got used to them. Perhaps the most troublesome is the difficulty of establishing a thorough audit trail of the sequence of events that led to the data within a particular record having the values that it has. Data Algebra can fix this problem. We will discuss how it does so in the next chapter, but before we go there, we need to draw attention to the reason why it took the computer industry so long to invent data algebra.

The Relational Database Malaise

As we noted in Chapter 2, the early work on relational database involved an attenpt to create a data algebra. However instead of producing a genuine data algebra, something that went by the name of "Relational Algebra" emerged. That was a misnomer as Relational Algebra was not at all a formal algebra. It was simply bad math.

The idea underlying relational database is to organize data into tables of rows and columns. In general, every entity (a thing: person, product, order, invoice, etc.) has its own table and the various attributes that describe it are columns in the table. Each row in the table represents one instance of the entity (one specific person or product or whatever).

In practice, not all tables represent entities, but those that do assign a unique key to each row. This is a physical implementation tactic that ensures the uniqueness of rows, and is useful for linking one entity to another (for instance linking a customer to an order).

All of this works effectively for data that fits well into tables. If relational database theory had confined itself to managing data in tables, then it is possible that a valid, but very limited, algebra could have been constructed around it. It could, perhaps, have based itself on matrix algebra (a matrix is a table or rectangular array of numbers, symbols, or expressions, arranged in rows and columns).

Unfortunately, that possibility was demolished by the introduction of SQL as a standard query language and, particularly, by the introduction of the null value, a wholly unmathematical idea, as we described in Chapter 2.

Good Math

In 2004, a company that went by the name of XPRADA Corporation, was founded in Austin, Texas with the stated mission "to produce a next-generation data management system grounded in advanced set theory." The President and COO was David L R Stein, a computer industry and start-up veteran who had earlier in his career been a cofounder of The Gartner Group. Christopher Piedmonte, a software engineer served as both CEO and CTO.

Piedmonte and Stein collaborated with Professor Gary J. Sherman, who had been Piedmonte's Math Professor at Rose-Hulman Institute of Technology, and Sherman led the research into the mathematical possibilities of applying set theory to the problems of data management. At first he investigated the Extended Set Theory (XST) of David L. Childs, who had proposed it as a database foundation in a 1968 paper and more prominently at the Very Large Data Bases Conference in 1977. Childs' idea was that set theory could provide a universal formal model for data storage and retrieval, an ambition E.F. Codd had partially shared when developing the relational model.

Sherman concluded that while Childs was correct in his aims, his "extension" to set theory was unnecessary. He instead focused on standard Zermelo-Fraenkel set theory as the foundation, and spent roughly four years hammering out a working algebra of data that could represent any data in any form.

Stein and Piedmonte's initial ambition to create a Data Algebra was realized by Professor Sherman, who then served from 2008 to 2014, as the company's Principal Mathematician. In 2009 the company changed its name to Algebraix Data Corporation and from 2010 through to 2017, Charles H. Silver, a serial entrepreneur who had previously founded and served as CEO of RealAge Inc. became involved initially as an investor and subsequently as Chairman and CEO.

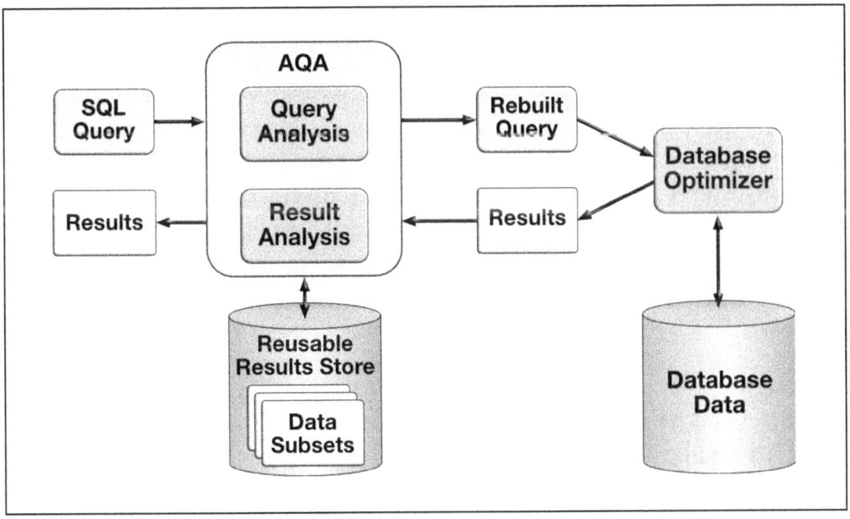

Figure 15 The Algebraix Query Accelerator

Under Silver's leadership, the company raised $45 million in funding (all from individual investors), grew its staff and developed database software.

Database Query Optimization

The primary purpose of a database is to provide swift access to data. Conceptually, a database is a shared pool of data that applications and database users can access concurrently. Various techniques are used to optimize the speed of retrieving data from disk. Such optimization techniques include:

- Caching: the holding of subsets of frequently used data in local memory so that it can be accessed faster.
- Indexing: indexes provide a fast mechanism for retrieving data.
- Data structures: data can be held in a convenient form that suits the query workload, such as storing data in columns.
- Data compression: compression reduces the space data items occupy, speeding up the movement of data from disk to memory or from memory to CPU.
- Query plan reuse: the caching of query plans in memory and their reuse.

These techniques are, to some degree, hardware agnostic. For example, replacing spinning disk with SSD does not alter the fact that SSD is slower than memory. As hardware evolves, database software adjusts to cater for the changes and might on occasion be extensively rewritten. However, the reality remains that a computer's resources have varying costs and speeds (CPU is faster than memory which is faster than storage technology). Once data expands beyond the capacity of a single server, network speeds are factored in. Every part of the hardware configuration has limited data capacity and a maximum speed. Database software works within these parameters.

Although the techniques may vary from product to product, all database optimizers take the same general approach. It can be described as follows: there is a regularly updated collection of data that users wish to query. The pattern of queries (normally SQL queries) is analyzed as data is served up and various techniques are employed that seek to access the data as fast as possible, while balancing the available resources between multiple, perhaps hundreds of concurrent queries. So the optimization techniques are based upon the relationship between the stream of queries and the pool of stored data they target.

AQA, The Algebraix Query Accelerator

Data algebra enables a new and novel approach to query optimization, one that complements the traditional approach. In its early years Algebraix Data built different database prototypes. Ultimately it focused on what it called the Algebraix Query Accelerator (AQA) — a system that translated queries into their algebraic form, analyzed them algebraically, and stored subsets of the results for possible reuse. The idea was that this software layer would not be software specific, but could be implemented over any data store.

The underlying principle was that most query traffic is repetitive, accessing the same data, performing the same table joins and producing results which overlap subsets of previous query results. AQA stored reusable results — not all results, just those partial results that were frequently calculated. The technique could be thought of as the application of heuristic intelligence to the database tuning role of the DBA. AQA learned more about the actual workload as it processed every additional query.

A simplified diagram of AQA's operation is provided in Figure 15. AQA intercepts all SQL traffic heading for the database. It parses each SQL query algebraically, transforming it into an algebraic representation. It holds a store of results and subsets of results from previous queries that it has processed– along with their algebraic representation, which are held in an algebraic catalog. To determine whether there is an opportunity to reuse any of the "chunks" of stored results, it consults its algebraic catalog, comparing the algebraic representation of the query with the catalog.

When it encounters a situation where some of the data it has stored can be reused, it reformats the SQL query to request only the data that is not held in the results store. It then passes that trimmed query to the database optimizer for resolution.

When the optimizer returns its results, the accelerator merges those results with the cached result it was reusing and returns the complete answer to the application that has issued the query. In this way, the database's workload is reduced and query performance improved.

In reality it doesn't matter whether the database that AQA sits on top of is highly performant or not, AQA will improve its performance. AQA also had the virtue that it could fit in front any data store that could accept SQL queries. It was implemented over a variety of relational databases and also over Apache Spark (with Spark SQL).

The Complementary Nature of AQA

Very few of the techniques that database optimizers perform automatically or that database administrators choose to implement duplicate the techniques that AQA employed. The one exception was materialized views. A materialized view is a database object containing the results of a query, used to pre-calculate common table joins or to precompute summaries that are regularly requested. Some databases (Oracle, DB2, Microsoft SQL Server and a few others) include this feature, but most do not. When available, it is an option a DBA can manually select to improve performance.

With AQA, the DBA no longer needed to consider using materialized views. AQA identified and created the effective ones automatically. In fact it created many such reusable results that no DBA would have identified as frequently occurring. Because of its algebraic analysis and the statistics it gathered, it could identify every candidate for reuse and store the ones that offered the greatest performance improvement. In general the most resource-intensive operations a database performs are table joins, sorts, summaries, and aggregations, and AQA significantly reduced this workload.

Federated Queries and the UQA Vision

Algebraix Data's longer-term intention was to evolve AQA to provide a capability similar in many respects to a federated database system — one that could query multiple database instances as if they were a single logical database. It referred to this vision as a Universal Query Accelerator (UQA), since it would not only federate queries across multiple databases but accelerate them.

Federating several databases is often a better option than trying to merge them. It eliminates the need to perform data transfers and transforms, and the federated databases remain current at all times. However, queries to federated databases are generally much slower than queries to large merged databases, because the federated query must be decomposed into subqueries — one for each physical database — and the results must be merged after the fact. The UQA would address this latency by maintaining a results cache across all the federated databases, drawing on the same algebraic acceleration principles that AQA employed on a single database.

The vision was expansive: multiple instances of the UQA, each presiding over several physical databases, communicating and sharing their data catalogs. In the long term, the UQA would perhaps span multiple data centers and reach into the cloud, becoming both a universal query accelerator and a global data directory.

Sadly, not of this came to pass. Despite having proved its unique acceleration capabilities in benchmarked envieronments and also having registered a whole sheaf of patents[1] to protect its unique capabilities, Algebraix data was not a commercial success.

In 2017, the company pivoted to pursue a blockchain-based business initiative and shuttered the remarkable database work it had done.

[1] *By 2016, the company had registered and been granted nine U.S. patents with 11 additional applications pending, plus foreign patents (filings in the EU, Australia, Canada, Japan, and China are documented in the patent records).*

The technology worked. The mathematics was sound, the performance improvements were real, and the principles of algebraic query acceleration and results-based optimization were validated in practice. But commercial success in the software world is an altogether different matter from technical validation.

Chapter 12: JabrDB, a Reunion of Broken Parts

"Whoever speaks of algebra, speaks of a universal art."

~ *Gottfried Leibniz*

Jabr

The word "algebra" entered European languages from the title of a ninth-century treatise by the Persian mathematician al-Khwārizmī, which went by the title, *al-Kitāb al-Mukhtasar fī Hisāb al-Jabr wal-Muqābalah*, which translates to: *The Compendious Book on Calculation by Completion and Balancing*. The *al-jabr* of the Arabic title referred to the process of moving a subtracted quantity to the other side of an equation, making it whole again. It meant, literally, "the reunion of broken parts."

As such Jabr is in our view a fitting name for a technology which enables a reunion of the fragmented empire of computer data. The world's data is splintered into billions of shards, scraps and tatters, scattered across file systems, databases, data lakes, cloud storage, email servers, and devices of every description. Each fragment is stored in its own contextual format, described (if described at all) by its own metadata and accessible only through its own chosen interface. The data archipelago is a vast topography of broken pieces. What data algebra offers, and what the Jabr technology embodies, is the mathematics that may one day reunite it all.

JabrDB- The Algebraic Database

It was not until 2025 that work began to build a truly algebraic database. It is possible that one was built during the prototyping work that Algebraix Data carried out, but if so it was only done in an effort to test and refine the query acceleration that was the primary target of that company. It was probably not considered likely that database users would want to learn data algebra and put it to use.

But when the author of this book was pondering the question of whether database users would be willing to adopt a different query language than SQL, he eventually concluded that an algebraic query language would be superior in many ways to SQL and with that in mind he began work on building an algebraic database. That was the genesis of JabrDB.

So, JabrDB is an algebraic database, built from the ground up on the mathematical foundations described and explained in this book. It should be noted, by the way, that the first nine chapters of this book were not intended to be exercises in data algebra. They were written as a roughly crafted mathematical proof that, using an algebra based on couplets and elaborating it to slightly more complex set-based structures, you could construct an algebra that would be able to represent all data, without exception.

Those first nine chapters have the spirit of a mathematical proof dressed up in a series of explanations. In this chapter we can begin to put the mathematics behind us. It has done its work. We can now discuss how to use algebraic tricks and techniques to usefully manipulate data.

Smooth Operators

We make our best effort here to describe the workings of JabrDB, but to be honest, you will find this chapter a great deal easier to navigate if you hold the book in one hand and the keyboard in the other. Get the app, it's not expensive.

You will know by now that an algebraic database like JabrDB stores data as hordes, clans, relations and couplets — not merely the rows and columns that haunt the relational database model - nor as the documents you might find in a document store - nor as the vertices and edges that inhabit a graph database. An algebraic database holds every data structure in its algebraic form and provides an algebraic query capability for accessing and manipulating that data.

In JabrDB, the fundamental query operations are the operations of data algebra itself. There are 23 operators, although many of them are similar versions of each other. And if you thnk that is a large number compared to SQL, actually it's not. SQL has roughly 50-60 operators and functions in the core standard, and possibly more depending on the SQL dialect you happen to abide by.

Also, although you might not appreaciate it until you use them, all the 23 operators have clean algebraic properties. They do what they do always and forever, in the same way, plain and simple.

Nevertheless the algebra seems a little crowded, almost two dozen operators, each with its own glyph — but almost none of them are independent. They fall into a small number of families, and within each family the members differ along just one or two simple axes: left versus right, keep versus drop, this level versus the one above. Once the families are visible, the apparent crowd collapses into roughly six ideas, repeated with variations.

The Tower

Data algebra inhabits a tower of four levels — couplets, relations, clans, hordes — and each level is, structurally, a set of the things composed from and standing on the level below. So any operator that works at one level reappears one level up, applied across the members. Thus a union becomes a cross-union which becomes a horde cross-union. It is not three operators; it is one operator at three altitudes.

We need to be aware of "handedness." A couplet has a left hand part and a right hand part: yin and yang. Several operators come in matched left/right pairs simply because they can act in two ways. That doubles the apparent operator count without introducing a single new idea. So, let's list every operator, organized into families by what it actually does. Incidentally, by listing the algebraic operators we are listing most of the verbs of JQL, the algebraic query language..

1. The set operators

Because every level of the tower is a set, ordinary set algebra applies, unchanged, everywhere.

∪	Union - includes everything in either operand.
∩	Intersection - includes only what is in both.
−	Difference - includes what is in the first but not the second.
⊕	Symmetric difference - includes what is in exactly one of the two.

These four are the same four whether the elements are couplets (relation level), relations (clan level), or clans (horde level). You learn them once.

Complement (′) is the near relative: difference taken against a fixed universe rather than against a second operand — "everything in the universe that isn't in R."

2. The selectors – pulling elements out of a relation

A relation is a set of couplets, each with a left and a right. These operators discard the structure and hand back a plain set of elements. The symbols we have chosen are our invention, for the sake of clarity.

 Domain (yin) – the set of all left elements.

 Range (yang) – the set of all right elements.

 Support – the union of domain with range; hence, every element that appears anywhere.

Domain and range are the genuine pair of operators, distinguished only by handedness. Support is their union.

3. Transpose – the mirror

↔ | Transpose — swap left and right in every couplet.

Transpose is the operator that relates the two selectors: the domain of a couplet transposed is the range of the couplet, and vice versa. Transpose twice and you are back where you started. Everything that comes in a left/right pair is really one operator seen through this mirror.

4. Composition and the diagonal – the multiplication

This is the one family of operators that is genuinely relational rather than set-theoretic. It is the engine of the algebra.

o | Composition – follows couplets of the first relation into couplets of the second where they meet: if (a, b) ∈ R and (b, c) ∈ S, then (a, c) ∈ R o S. This is "join the right of one to the left of the next," and it is how chains and paths are expressed.

δ | Diagonal – this is the set of couplets (a, a) for each element a. It is the identity element for composition: compose anything with the diagonal and you get it back unchanged, exactly the same as multiplying by one.

Composition and the diagonal belong together the way multiplication and the number one belong together.

5. The restriction family – restricting one relation by another

A clean two-by-two. Each operator keeps or drops couplets of the first relation according to whether one of their elements appears in the domain (or range) of the second.

▷	Superstriction – keep the couplets of R whose left element is in the domain of S.
◁	Substriction – keep the couplets of R whose left element is not in the domain of S.
▶	Right-superstriction – keep the couplets of R whose right element is in the range of S.
◀	Right-substriction – keep the couplets whose right element is not in the range of S.

Just one simple idea – filter one relation by another's elements – along two axes: in or out, and left or right. This family is where selection lives (the SQL WHERE clause, roughly).

6. Lifting – the cross operators

The cross operators increase the operator count dramatically, but in reality this is the simplest idea of all. We carry a binary operator up one level of the tower, applying it to every pair of members. A clan is a set of relations, so the cross-version of an operator runs it over all pairs of relations – one from each clan – and collects the results. The symbols we use follow the same pattern, surrounding the familiar operator with a broken circle.

(∪) (∩) ⊖ ⊕	All the set operators lifted: Cross-union Cross-intersection Cross-difference Cross-symmetric-difference
⊚	Cross-composition – composition, lifted.
(▷) (◁) (▶) (◀)	Cross-super/substriction (and their right variants) – the whole of the restriction family, lifted.

Nothing here is new. Each cross-operator is its level-below counterpart wrapped in "do it for all pairs." And the same wrapping happens again from clans to hordes — horde cross-union, horde cross-composition, and so on. One mechanism, applied twice up the tower.

7. Natural join – the famous one, and it is derived

$\bowtie$	Natural join – match relations on their shared structure and merge them.

The natural join is an operator everyone should recognize from relational databases, but in this algebra it is not primitive. By that, we mean we didn't need to include it as it can be assembled from the cross-operations we previously described. It takes its place primarily because it will be familiar to most database users and because it is the workhorse of the clan level. Unusually among the join operators — it is commutative.

8. The structural operators – reshaping a table

The families so far act on the contents of relations and clans. This last group acts on their shape — which columns exist, and how rows are grouped.

π	Project – keep only certain columns and drop the rest.
ρ	Rename – change column labels without touching the data.

These are the operators that let the user make a clan behave like a table you can query and report on.

9. The odd man out – the complement

Last, and to be honest, in my view, also least (but not zero) is the complement operator.

$'$	The complement turns a relation inside out: it gives you everything that isn't there.

A relation is a set of couplets. To take its complement you first fix a universe — the full set of couplets you're considering possible — and then the complement of R, written R′, is everything in that universe that R leaves out. So R′ = universe – R.

Clearly this is most useful when your universe is small. Say {(a, a), (a, b), (b, a), (b, b)}. So if R = {(a, a) ,(a, b)}, R′ ={(b, a), (b, b)}.

JabrDB

Here we describe the current operation of the JabrDB database, noting that what we describe is the first version, implemented on Apple Mac. The app, which may have changed to some degree, by the time you read this book. We intend to produce a Windows version of the app as soon as possible after the release of the Apple Mac app, and subsequently a Linux version.

(You will find this part of the book easier to understand if you simply acquire the app and experiment with it. It is priced inexpensively.)

All of the illustrations we show below are screenshots taken directly from the running app.

JabrDB — The Window and its Four Regions

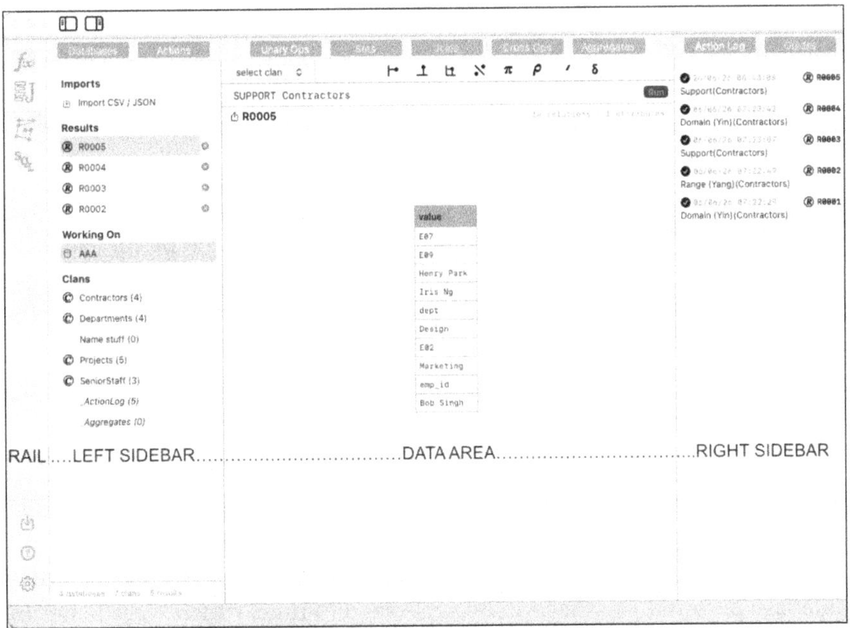

Illustration 1. The JabrDB app window.

When JabrDB opens a database, the work happens inside a single window divided into four regions as labeled in the above illustration: The Navigation Rail, the Left Sidebar, the Data Area, and the Right Sidebar. The division is not arbitrary; each region answers a different question. The rail decides what kind of work you are doing; the Left Sidebar decides which data you are working on; the Data Area is where the work is done and data is displayed; and the Right Sidebar keeps a log, a record of what you have done– it also provides a guide for new users of the app. Once that logic is clear, the window explains itself.

The two sidebars can each be folded away using the small toggle controls at the left end of the title bar. This can be useful when the user wishes to display as much information as possible from a clan. And the window itsefl can be exapnded to become as large as the display allows.

The Navigation Rail

The rail is the narrow vertical strip down the left edge of the window — the spine of the application. It is the place where different application modes reside. In its first release, JabrDB will provide an algebraic database. Additionally, or soon thereafter, it will provide a forms capability that allows the user to set up forms for data entry into the algebraic database.

Illustration 1 also shows two other icons in the Rail, separated by a divider from the first two. These are the icons for a Graph database and a SQL database.

Choosing one of these on the Rail puts you into a different database capability, where you can create or manipulate corresponding types of database structure. In practice you are switching to a new type of database and application, but as the user interface is very similar between these different database personalities, it all resides within a single application. This is possible because data algebra can represent any kind of data, and hence can, in its own way, cater for any kind of database application

The Graph database and SQL database are planned additions to the JabrDB portfolio of database "personalities." Once these databases are available and if the user purchases them, then clicking on them will initiate a graph database capability and a SQL database capability. There will also be other database personalities that will be supported in time, which we list later in this chapter..

The three icons in the bottom area of the Rail, are respectively:

1. An Import capability, that allows CSV or Json files to be imported as clans.
2. A help capability, which thoroughly documents the application.

3. A settings capability that allows the user to change some of the basic parameters of the application.

The Left Sidebar

The Left Sidebar is the catalog. It has two tabs across its top — Databases and Actions — and the tab you choose changes what the sidebar lists.

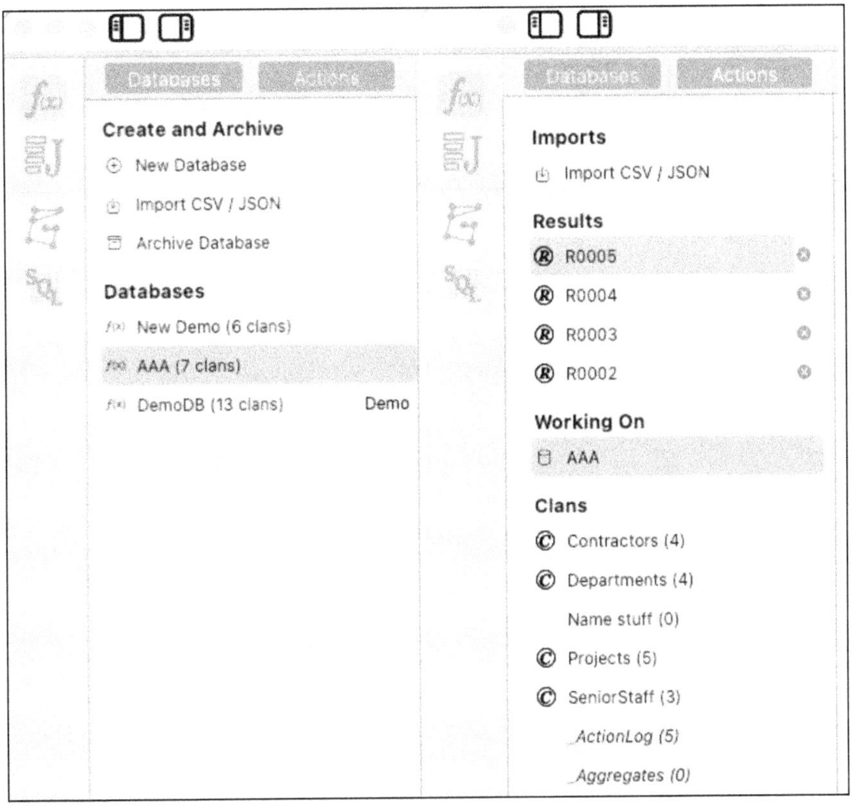

Illustration 2. Two Left Sidebar (Databases active and Actions active)

On the left side of the illustration we show the Databases tab. It lists every database JabrDB can see. Because a database in JabrDB is simply a folder of files, this is in practice a list of those folders, each shown with an icon, the number of clans it contains, and, where relevant, a small Demo marker to indicate the sample database that ships with the application. Selecting a database here opens it, making it the active database. It is then possible to click on the Actions tab.

The Actions tab turns to the database that is currently open and lists the clans inside it — JabrDB's term for a table. Each clan name carries a small circular badge that records where the clan came from: C for a clan imported from a CSV file, J for one imported from JSON, R for one produced as the result of an operation, and B for one built by hand in JustForms.

Two clans always sit at the foot of the database's clan list, their names shown in italic and without a badge. They are the Action Log and the Aggregates clan, both of which are automatically maintained by the application itself. The same tab also surfaces your most recent imports and your most recent result clans, so the things you are most likely to reach for are never more than a glance away. Selecting any clan displays it in the Data Area.

The Data Area

The Data Area is the center of the window and the place the real work happens. It is organized top to bottom as a workbench.

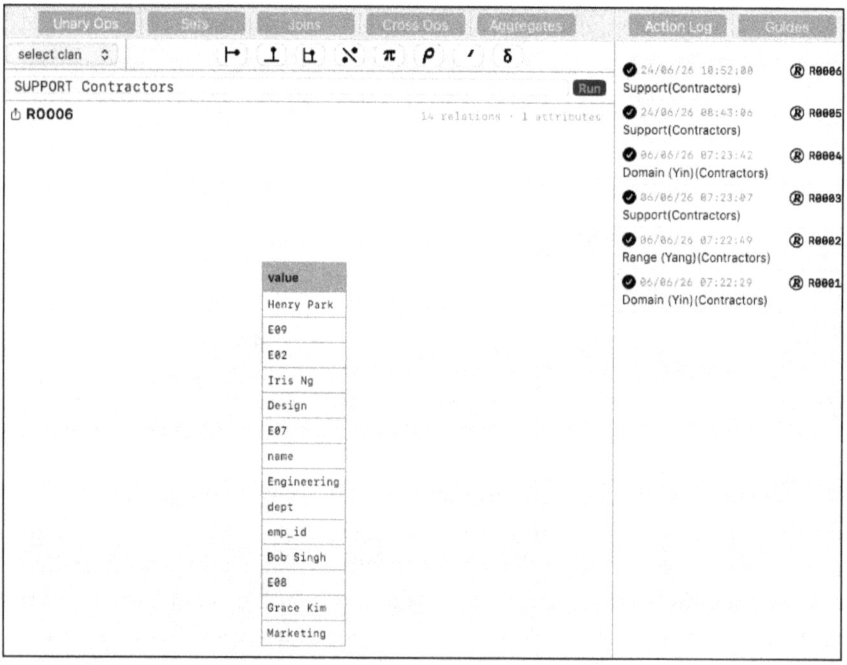

Across the top runs a tab bar offering the five families of operation: Unary Ops, Sets, Joins, Cross Ops, and Aggregates. Choosing a tab determines which operators are available below it. Beneath the tabs sits the console, the strip where an operation is assembled: you pick a clan to work on, pick a column where the operation needs one, and pick the operator itself from a row of buttons. As you make those choices, the application writes the corresponding JQL — JabrDB's query language — into a statement line, so the console doubles as a teacher: you can watch the algebra take shape in words even if you never type a character of it. A Run button, drawn as a colored pill, executes what you have composed.

The result appears immediately below, in the Results Area, as a clan in its own right — a named heading followed by a table of rows and columns. It is also listed in the Action Log in the Right Sidebar. Every result is kept: it is given an automatic name in the series R0001, R0002, and so on, registered in the workspace, and made available as the input to the next operation, so a sequence of operations can be chained into a pipeline.

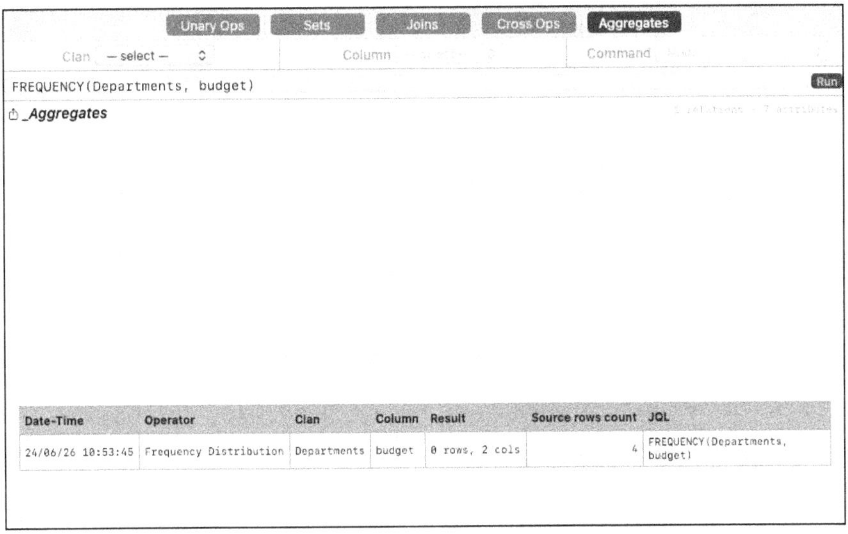

Illustration 4. The Data Area after an Aggregate calculation

Illustration 4 shows the result of an aggregate calculation. When the user runs an aggregate calculation by selecting a clan, then a column, then a type of aggregation, the JQL statement is shown and when run it provides a result similar to the one shown, which is a frequency calculation. This immediately becomes an entry in the *_Aggregates* clan, which, over time, becomes the record of aggregates at specific dates and times.

Right Sidebar

The Right Sidebar is the window's memory and its guide. Like the Left Sidebar it has two tabs — Action Log and Guides.

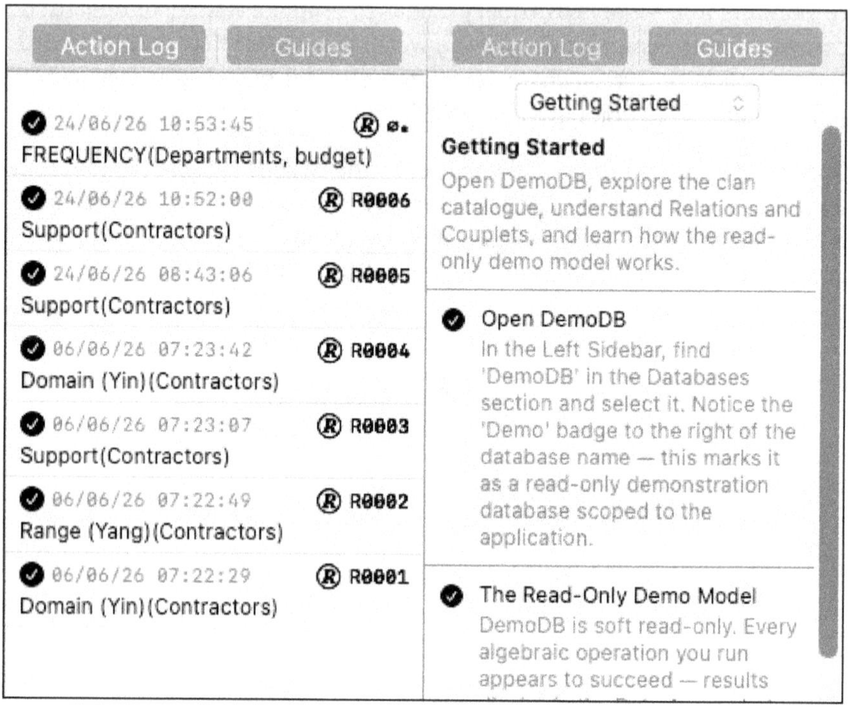

Illustration 5. Two views of the Right Sidebar

On the left we show the Action Log. This is an append-only record of everything the user has done in the open database, listed newest first. Each entry notes the operation that ran and carries a clickable result-code link — an Rnnnn — that jumps straight to the result that operation produced. Nothing here can be edited or individually deleted; it is an audit trail, the running history of the session, and the companion to the Action Log clan kept in the Left Sidebar's list. When the log is empty, the area simply says so.

On the right we show the Guides tab. This is only usable when the demo database is selected. The guides are short, structured tutorials. At the top is a dropdown list of available guides. When one is chosen, its full text is displayed in a scrollable panel. The set of guides offered is sensitive to what the user is looking at, so the help on hand tends to match the work in front of the user. The Right Sidebar, like its counterpart on the left, can be folded shut when you need the space.

How the four regions work together

The flow across the window is left to right and back again. You choose a mode on the rail; you pick the database and clan to work on in the Left Sidebar; you compose and run the operation in the Data Area and read its result there; and the Right Sidebar records what you did and offers guidance as you go. Each region has one job, and because every result is itself a clan that re-enters the Left Sidebar's list and the Data Area's selectors, the four regions form a loop: the output of one piece of work becomes the raw material of the next.

The Database Personalities

As already noted there are a number of different database personalities that can be added on top of the JabrDB Algebraic database. They are as follows:

JabrDBS (SQL) — SQL keyword aliases: CREATE TABLE, INSERT, UPDATE, DELETE, SELECT...FROM...WHERE, ORDER BY, HAVING, LIMIT. Each maps to an algebraic equivalent (SELECT→project, WHERE→superstrict, JOIN→compose/natural join, etc.). These are syntactic sugar, not new operations.

JabrDBG (Graph) — ADD TRIPLE, QUERY (SPARQL-style pattern matching), GRAPH, NAMESPACE, PREFIX. Triples are Couplets; graph traversal is iterated composition; SPARQL patterns are superstriction against pattern Relations.

JabrDBV (Text) — Combination of XML, Vector and Full-Text Search: ADD VECTOR, SEARCH (k-NN), SIMILARITY, DISTANCE, INDEX. This introduces a genuine new computational capability: distance metrics and nearest-neighbor search. A standard function library for numeric vector operations is included.
XML: XPATH, XQUERY, VALIDATE, TRANSFORM. XML documents map to nested Couplets/Relations. XPATH navigation is composition along nested structure. VALIDATE and TRANSFORM are constraint-checking and restructuring operations expressible as superstriction and projection over nested structures.

JabrDBM (Media) — TAG, SEARCH MEDIA, CATALOG, METADATA, COLLECTION. Media metadata is a Clan; tagging is union of a tag Couplet into a Relation; search is superstriction against tag/metadata filters; collections are named Clans.

JabrDBO (OLAP / Cube) — SLICE, DICE, ROLLUP, DRILLDOWN, PIVOT, AGGREGATE. A cube is a Clan of cells, each cell a couplet from an n-tuple of dimension coordinates to a measure (the measure is the cell's yang). Each dimension is itself a Clan; its hierarchy is a parent-child Relation L, so ROLLUP climbs the levels by composition (L, $L \circ L$, ...) and DRILLDOWN descends via the transpose of L. SLICE and DICE are superstrictions that pin the fact Clan to fixed members; AGGREGATE partitions by a level and folds the measure; PIVOT reorients the axes — transpose ($\leftrightarrow$) in the two-dimensional case, a permutation of the coordinate n-tuple in general.

JabrDBK (KeyValue) — GET, PUT, DELETE, SCAN. The store is a single yin-functional Relation $R \subseteq G \times G$: keys are the domain (yin), values the range (yang), and the functional predicate is what guarantees one value per key. GET is superstriction by a singleton key followed by yang; PUT overwrites by substricting the old key then unioning the new couplet; DELETE is substriction by the key set; SCAN over a key range is superstriction to a key subset. Multiple namespaces or buckets are a Clan of such Relations.

JabrDBC (Cartographic) — LOCATE, REGION, NEAREST, OVERLAY, PROJECT. A layer is a Clan of features, each feature a couplet (id $\rightarrow$ geometry); a full map is a Horde of layers, exactly as the medallion is a Horde of zones. Spatial indexing is a Relation T mapping tiles to feature ids, so a REGION query superstricts T to the tiles covering the window and then composes to recover the features. OVERLAY of two layers is cross-intersection; topological adjacency between regions is a Relation A with reachability $A \circ A$; NEAREST and point-in-polygon are natural joins ($\bowtie$) across layers; PROJECT is composition with a coordinate-transform Relation.

JabrDBL (DataLake) — INGEST, PROMOTE, PROFILE, ZONE (bronze/silver/gold). The medallion architecture is modeled as a Horde with three named Clans (zones). PROMOTE moves Relations between zones via substriction from source and union into target. PROFILE is a derived Clan of statistical metadata.

JabrDBT (Time Series) — The largest personality-specific extension: TIME_BUCKET, DOWNSAMPLE, UPSAMPLE, SAMPLE_BY, GAP_FILL, INTERPOLATE, LOCF, MOVING_AVG, EWMA, LAG, LEAD, DIFF, PCT_CHANGE, CUMULATIVE, TWA, RATE, ASOF_JOIN, WINDOW_JOIN, OHLC, VWAP, SPREAD, ZSCORE, ANOMALIES, FORECAST, SEASONAL, ROLLUP, RETAIN, LATEST, RANGE, SINCE, FIRST_N, LAST_N. Structurally, time-series data is a Clan with a timestamp yin. Most of these are aggregation and window functions over partitioned Clans (Hordes). ASOF_JOIN is composition with nearest-match semantics. The statistical and forecasting functions (EWMA, FORECAST, SEASONAL, ZSCORE) require a numeric function library beyond pure set algebra.

This is not the full complement of the various types of database that have been investigated, but it is a list of those where we intend to build products as we move forward.

Chapter 13: The Sorry Saga of SQL

When the world has once begun to use us ill, it afterwards continues the same treatment with less scruple or ceremony, as men do to a whore.

~ Jonathan Swift

SQL–and its Sequels

It might seem that we are taking an alarming digression here, by suddenly turning our attention to SQL, but we are not. It is the gorilla in the room and it needs to be discussed.

Cast your mind back, if it stretches that far back, to the days before the invention of SQL. There were a number of very capable databases. Aside from IBM's own IMS/DB, there were IDMS, ADABAS, TOTAL, Datacom/DB, and Model 204, and incidentally, all these products are still running in various necks of the woods. However the computer scientists of the day were well aware that all of these excellent products had their own data access methods. Would it not be a wonderful thing if there was but one access method defined perhaps in a query language that could be used for all databases no matter how they physically arranged the data on spinning disks.

And so it came to pass in the early 1970s that Donald D. Chamberlin and Raymond F. Boyce at IBM's San Jose Research Laboratory invented a data access language that they called SEQUEL (Structured English QUEry Language). They had learned about Ted Codd's relational model and set out to create a practical language for manipulating and retrieving data stored in System R, IBM's prototype relational database. The name SEQUEL was later shortened to SQL for trademark reasons.

Through the late 1970s and early 1980s, SQL spread beyond IBM. Oracle (then Relational Software Inc.) shipped the first commercial SQL-based product in 1979. IBM followed with SQL/DS in 1981 and DB2 in 1983. By the mid-1980s, multiple vendors were shipping SQL databases, each with its own syntax and semantics.

SQL Standards

Well, of course, that wasn't good. And thus the gods of computing decreed that there should be a standards committee. And thus the Accredited Standards Committee X3 (now INCITS) developed the first SQL standard, which was adopted by ANSI in 1986. ISO published a technically identical standard, ISO 9075-1987, the following year. What follows documents the evolution of SQL. Read this only if you're a standards fanatic.

SQL-86

SQL-86 as it came to be known codified the core SELECT/INSERT/UPDATE/DELETE syntax and defined bindings for four host languages: COBOL, FORTRAN, Pascal, and PL/I. It was largely a formalization of the IBM dialect as it was. And that needed some improving.

SQL-89

The first enhancement, SQL-89 (ANSI X3.135-1989 / ISO/IEC 9075:1989) was a minor revision. The principal addition was the inclusion of referential integrity constraints (FOREIGN KEY, REFERENCES). The standard remained a single document.

SQL-92

Thence came SQL-92 (ANSI X3.135-1992 / ISO/IEC 9075:1992) which was a major expansion. It tripled the size of the standard and introduced three conformance levels — Entry, Intermediate, and Full — presciently acknowledging that no vendor was likely to implement the whole thing. The key additions included: JOIN syntax (INNER, LEFT/RIGHT/FULL OUTER, CROSS), CASE expressions, CAST for type conversion, string operations (SUBSTRING, TRIM, concatenation with ||), dynamic SQL, temporary tables, scrollable cursors, and the INFORMATION_SCHEMA.

In practice, SQL-92 Entry Level became the de facto baseline that most vendors eventually claimed to support. Time passed, the millenium was approaching and it was clearly necessary to get a new SQL standard into existence before all the computers in the world failed to fail on 1st January 2000.

SQL:1999

That was SQL:1999 (ISO/IEC 9075:1999). Note here, by the way the radical naming adjustment. They replaced the dash of SQL-92 with a colon as in SQL:1999. This was a substantial expansion. It introduced: regular expression matching, recursive queries (WITH RECURSIVE and Common Table Expressions), triggers, user-defined types (structured types, distinct types), roles for access control, OLAP functions (ROLLUP, CUBE, GROUPING SETS), savepoints, and the procedural language framework (SQL/PSM — Persistent Stored Modules). The three-level conformance structure was superseded by a Core SQL specification plus optional feature packages.

SQL:2003

The fourth in line revision to SQL was SQL:2003 (ISO/IEC 9075:2003). This added window functions (OVER, PARTITION BY, ROW_NUMBER, RANK, DENSE_RANK, LEAD, LAG), the MERGE statement (upsert), sequence generators, XML-related specifications (SQL/XML as Part 14), multiset types, BIGINT and BOOLEAN data types, and the TABLESAMPLE clause. Window functions were arguably the most consequential addition since SQL-92's JOIN syntax.

SQL:2006

The fifth revision was SQL:2006 (ISO/IEC 9075:2006). This was a focused revision that substantially expanded SQL/XML — defining how to import, store, query (via XQuery), and export XML data within SQL databases. Happily, there were no major changes to other parts of the standard.

SQL:2008

The sixth was SQL:2008 (ISO/IEC 9075:2008), a relatively modest revision: INSTEAD OF triggers, the TRUNCATE TABLE statement, enhanced FETCH clause (FETCH FIRST n ROWS ONLY as the standard pagination syntax), and various corrections to the existing standard.

SQL:2011

The seventh was SQL:2011 (ISO/IEC 9075:2011). In this we encounter temporal tables — system-versioned tables that automatically track row history, with temporal query syntax (FOR SYSTEM_TIME AS OF, BETWEEN, FROM...TO). Also added: pipelined DML, enhanced window functions (CUME_DIST, PERCENT_RANK, NTH_VALUE), and improvements to the CALL statement.

SQL:2016

And then there were eight. SQL:2016 (ISO/IEC 9075:2016) introduced 44 new optional features, dominated by two areas. First, JSON support: functions to create, access, and validate JSON data (JSON_VALUE, JSON_QUERY, JSON_TABLE, JSON_OBJECT, JSON_ARRAY, IS JSON). Second, row pattern recognition (MATCH_RECOGNIZE) for matching sequences of rows against regular expression patterns. Also added: polymorphic table functions, LISTAGG for string aggregation, and various date/time enhancements.

SQL:2023

And finally, although probably not that finally, we arrived at SQL:2023 (ISO/IEC 9075:2023). The most significant addition was a new Part 16: Property Graph Queries (SQL/PGQ), which adds graph query capabilities on top of existing SQL tables — defining graph patterns, path-finding, and graph traversal within the SQL framework. Also added: enhancements to JSON support (JSON_SERIALIZE, simplified JSON syntax), the ANY_VALUE aggregate function, and various corrections throughout.

This brief history tells its own story. The SQL standard has grown from what began as a single concise document in 1986 to a multi-part specification estimated at over 4000 pages - equivalent to roughly 10 substantial novels. Each revision has added new syntax to handle data shapes and query patterns that the original design did not anticipate: recursive queries, XML, JSON, temporal data, window analytics, row pattern matching, graph traversal.

The language has grown because the underlying data model (flat tables of typed tuples) cannot accommodate new requirements without new syntax.

And despite nearly four decades of standardization, no vendor fully implements the SQL standard, and every significant vendor has added proprietary (i.e. non-standard) extensions that their users depend on.

Below is a catalog of the principal SQL dialects, grouped into eight different categories. We do not expect you to read all the entries, maybe just browse them a little. So How Many?

It is difficult to arrive at an exact count of SQLs. It depends on how you draw the boundary between "distinct dialect" and "minor fork." A reasonable tally of actively maintained, materially distinct SQL dialects gives a range somewhere between 40 and 50.

SQLFluff, the SQL linter (static analysis tool that checks SQL code for errors, style violations, and potential problems without actually executing it) currently supports about 30 dialect configurations. Apache Calcite, the SQL framework, defines dialect output for roughly 20 databases. Google Looker supports connections to over 50 SQL-based data sources.

If you include historical and discontinued dialects such as Sybase SQL Anywhere, Netezza SQL, Greenplum SQL, Vertica SQL, ParAccel, Teradata Aster, Pivotal HAWQ, and others, it's 60+. And it's a moving target. Each new cloud data warehouse, streaming engine, and embedded analytical database usually introduces its own SQL variant.

For the record, the following tables provide a fairly comprehensive picture of the current situation as of June (2026).

Traditional Enterprise RDBMS		
Vendor	**Dialect Name**	**Notable divergence**
Oracle	PL/SQL	Proprietary procedural language, CONNECT BY for hierarchies, ROWNUM pagination, NVL instead of COALESCE, (+) outer join syntax, object-relational extensions
Microsoft	T-SQL	TOP for pagination, square bracket quoting, + for string concatenation, proprietary date functions (GETDATE, DATEDIFF), IDENTITY columns, cross-apply
IBM	DB2 SQL	Closest to the standard historically, but has its own procedural extensions, OLAP syntax variations, MQT (Materialized Query Tables), XML extensions
SAP/Sybase	T-SQL variant	Shares ancestry with Microsoft T-SQL (both descend from Sybase SQL Server), but has diverged over time

Traditional Enterprise RDBMS		
Vendor	**Dialect Name**	**Notable divergence**
SAP	SAP HANA SQL	Column-oriented extensions, in-memory-specific syntax, calculation views, spatial extensions
Teradata	Teradata SQL	QUALIFY clause (filtering window function results), proprietary date arithmetic, COLLECT STATISTICS, hash distribution syntax
Informix	Informix SQL	BM acquisition; its own procedural extensions (SPL), collection types, time-series extensions
Open Source RDBMS		
Vendor	**Dialect Name**	**Notable divergence**
PostgreSQL	PL/pgSQL	Generally closest to the standard, but: folds unquoted identifiers to lowercase (standard says uppercase), SERIAL for auto-increment, rich extension ecosystem, array types, JSONB, custom operators, DO blocks
MySQL	SQL	Backtick quoting, LIMIT for pagination, AUTO_INCREMENT, non-standard GROUP BY behavior (historically), IF()/IFNULL(), ENGINE= storage engine selection
MariaDB	MariaDB SQL	MySQL fork with additional divergences: sequences, system-versioned tables, Oracle compatibility mode, window functions added earlier than MySQL
SQLite	SQLite SQL	Type affinity instead of strict types, no ALTER TABLE (beyond basic ADD COLUMN), manifest typing, no RIGHT/FULL OUTER JOIN (until 3.39.0), no stored procedures
CockroachDB	CockroachDB SQL	PostgreSQL-compatible wire protocol but distributed SQL with its own extensions for partitioning, locality, and serializable isolation

Open Source RDBMS		
Vendor	**Dialect Name**	**Notable divergence**
YugabyteDB	YSQL	PostgreSQL-compatible dialect for distributed SQL, plus its YCQL (Cassandra-compatible) interface
TiDB	TiDB SQL	MySQL-compatible dialect for distributed SQL, with divergences in transaction handling and some function behavior
Cloud Data Warehouses		
Vendor	**Dialect Name**	**Notable divergence**
Google	BigQuery SQL	STRUCT and ARRAY as first-class types, UNNEST for array flattening, backtick quoting for projects/datasets, scripting syntax, ML integration (BQML), no UPDATE/DELETE without WHERE, federated query syntax
Snowflake	SQL	VARIANT type for semi-structured data, FLATTEN for nested data, QUALIFY clause, time travel (AT/BEFORE), zero-copy cloning syntax, Snowpark extensions
Amazon	Redshift SQL	PostgreSQL 8.x derived, but: no arrays, limited JSON support, DISTKEY/SORTKEY declarations, WLM queue syntax, COPY/UNLOAD for bulk operations
Databricks	Databricks SQL	Spark SQL heritage, Delta Lake extensions (MERGE, time travel, OPTIMIZE, VACUUM), Unity Catalog syntax, ML functions
Microsoft	Azure Synapse SQL	T-SQL variant with distribution syntax (HASH, ROUND_ROBIN, REPLICATE), external table syntax, PolyBase extensions
Firebolt	Firebolt SQL	PostgreSQL-influenced but with its own aggregating index syntax, UNNEST, and engine management DDL

Streaming and Real Time		
Vendor	**Dialect Name**	**Notable divergence**
Apache Flink	Flink SQL	Streaming extensions: watermarks, event-time windowing (TUMBLE, HOP, SESSION), temporal joins, MATCH_RECOGNIZE, connector DDL
Apache Kafka	KSQL/ksqlDB	Stream/table duality syntax, EMIT CHANGES, window types, persistent queries, connector integration
Materialize	Materialize SQL	PostgreSQL-compatible with streaming extensions: CREATE MATERIALIZED SOURCE, temporal filters, tail queries
RisingWave	RisingWave SQL	PostgreSQL-compatible streaming SQL with CREATE SOURCE, CREATE SINK, watermark syntax
Embedded and Edge		
Vendor	**Dialect Name**	**Notable divergence**
SQLite	SQLite SQL	Dominant embedded SQL dialect
DuckDB	DuckDB SQL	PostgreSQL-influenced analytical dialect: list/struct types, PIVOT/UNPIVOT, friendly SQL extensions (FROM-first queries, column aliases in WHERE), in-process execution
Apache Derby	Derby SQL	Java-embedded, subset of SQL standard with Java stored procedures
New SQL and Distributed		
Vendor	**Dialect Name**	**Notable divergence**
Google Spanner	GoogleSQL	Globally-distributed, strongly-consistent; INTERLEAVE IN PARENT for table hierarchies, ARRAY/STRUCT types, stale reads syntax
VoltDB	VoltDB SQL	Subset of SQL with stored procedure focus, partitioned table syntax
SingleStore (MemSQL)	SingleStore SQL	MySQL-compatible with columnstore extensions, PIPELINE syntax for ingestion, JSON support

Time-Series and Specialized		
Vendor	**Dialect Name**	**Notable divergence**
TimescaleDB	TimescaleDB SQL	PostgreSQL extension with hypertable syntax, continuous aggregates, time_bucket functions
QuestDB	QuestDB SQL	PostgreSQL wire protocol with SAMPLE BY for time-series downsampling, LATEST ON for deduplication
ClickHouse	ClickHouse SQL	PostgreSQL-compatible with streaming extensions: CREATE MATERIALIZED SOURCE, temporal filters, tail queries
RisingWave	RisingWave SQL	Column-oriented analytical dialect: ENGINE= table engines, MATERIALIZED/ALIAS columns, approximate aggregate functions, array functions, FINAL keyword
Apache Druid	Druid SQL	Read-only analytical SQL translated to native Druid queries; limited DML
Legacy and Niche		
Vendor	**Dialect Name**	**Notable divergence**
Fujitsu	Symfoware SQL	Primarily Japanese market; proprietary extensions
Progress	OpenEdge SQL	ABL integration, proprietary syntax
InterSystems	IRIS SQL / Caché SQL	Object-relational with ObjectScript integration
Actian	Ingres SQL / Vector SQL	Codd-era heritage; QUEL predecessor
Mimer	Mimer SQL	Claims highest SQL standard conformance
ADABAS/ Natural	SQL gateway	Mainframe-era data with SQL access layer
Firebird	Firebird SQL	InterBase fork; EXECUTE BLOCK, generators, domains
H2	H2 SQL	Java embedded; multiple compatibility modes (PostgreSQL, MySQL, Oracle, SQL Server)
HSQLDB	HyperSQL	Java embedded; multiple SQL standard conformance levels

A Modus Operandi

The proliferation of SQL dialects is not accidental. It is a natural consequence of SQL itself. The concept of a single query language was from the get-go a marketing fantasy wrapped in a pipe dream.

Because SQL is a programming language, every genuinely new capability inevitably requires new syntax. Each vendor, pursuing its desired and often innovative direction and riding on its own technical architecture, invents an appropriate syntax to satisfy the user requests and naturally does so without considering the interests of any other SQL users.

Some time later, the standards committee meets and attempts to harmonize the proliferating dialects as best it can. But the committee has no teeth. If it "rules against" a particular innovation the vendor will most likely ignore it, after all the vendor will have an installed base that is already using the new syntax. And, in truth, it is the job of database vendors to innovate and compete.

Chapter 14: JQL v SQL

"You never change anything by fighting it; you change things by making them obsolete through superior technology."

~ Buckminster Fuller

JQL — The Full Monty

In this penultimate chapter we intend first to formally describe JQL, the JabrDB query language, so that the reader has a reasonable grasp of what it is and how it works. We then proceed to compare it to SQL identifying strengths and weaknesses.

Every database language is really several languages wearing a single name. The relational tradition, following the evolution of SQL, divides its vocabulary into five sublanguages::

1. A Data Query Language for reading data.
2. A Data Definition Language for declaring structure.
3. A Data Manipulation Language for inserting and changing rows.
4. A Data Control Language for granting and revoking access, and
5. A Transaction Control Language for committing and rolling back transactions.

This division is practical and useful — it tells newcomers which part of the language to reach for. So we will use the same approach here to lay out JQL in full.

Nevertheless, the above division is a child of the relational model, and JQL has no such parent.

JQL is the child of the algebraic tower, as illustrated in the table below:

Level	Structure	Definition	Intuition
0	Couplet	An ordered pair {a, b} drawn from Γ x Γ	A single fact: column → value
1	Relation	A set of Couplets – P(Γ x Γ)	A record (a row)
2	Clan	A set of Relations – P^2(Γ x Γ)	A table
3	Horde	A set of Clans – P^3(Γ x Γ)	A database, or a set of tables

When those five headings are laid over that tower, their proportions change and several of their borders dissolve.

In SQL the five sublanguages are roughly co-equal, each a substantial body of syntax. In JQL one of them, the query language, is almost the whole thing; the other four are thin layers defined in terms of it.

For this reason we begin by defining the data query language (DQL) and only then turn our attention to the rest. In JQL, everything else is built on the read algebra. Once that algebra is understood, definition, manipulation, control, and transactions follow with little ceremony.

1. Data Query Language (DQL)

Where SQL's query language manipulates tables into other tables, JQL's query language manipulates the four structures of the algebraic tower into other structures of the tower, using a small, closed set of operations that compose without restriction.

The Tower

Every value JQL reads or produces sits at one of the four levels, each of which is built from the level below:

The single most important property of the tower is that every operation defined at one level lifts to the levels above it. This is why JQL needs perhaps a dozen operations where SQL needs scores of keywords: the same operation, defined once on Couplets or Relations, is reused at every altitude.

JQL's operations fall into a handful of families.

Boolean operations

The Boolean operations treat structures as the sets they are, and work unchanged at every level: union, intersection, difference, symmetric difference, and complement (relative to a chosen universe). You learn these four-and-a-bit operations once and apply them to Couplets, Relations, Clans, or Hordes alike.

Structural operations

The structural operations reshape a Relation rather than combining two of them. Transpose swaps the left and right of every Couplet. Composition chains two Relations through their shared values — if (a, b) is in R and (b, c) is in S, then (a, c) is in $R \circ S$ — and it is the engine of reachability, paths, and, as we shall see, projection. Yin (yin(R)) returns the set of left elements — the column headers; yang (yang(R)) returns the set of right elements — the cell values; support (support(R)) is their union. The diagonal (D_G) is the identity Relation $\{(a, a) : a \in G\}$, the "one" that composition multiplies by. And rename relabels selected yin symbols, leaving every value untouched.

Restriction operations

The restriction operations are JQL's filter, and they are far more general than SQL's WHERE. Superstriction of R_1 by R_2 keeps the Couplets of R_1 whose left element appears in yin(R_2); substriction keeps those whose left element does not; the right-handed variants test the right element against yang(R_2) instead. The crucial difference from SQL is that the thing doing the filtering is itself a Relation — a first-class data object that can be stored, composed, and queried — not a string predicate trapped inside a clause.

Cross operations

The cross operations are how a binary operation climbs the tower. To lift an operation from Relations to Clans, apply it to every pair drawn from the two Clans: cross-union(C_1, C_2) = $\{ R_1 \cup R_2 : R_1 \in C_1, R_2 \in C_2 \}$. Every Boolean and restriction operation has a cross variant, and cross-union is how JQL performs joins. The natural join ($\bowtie$) is exactly a cross-union restricted to the pairs that agree on their shared columns — SQL's NATURAL JOIN, recovered algebraically, with no ON clause because agreement on shared yin is the join condition by definition.

How familiar queries are expressed

Because the operations compose freely, the everyday repertoire of SQL falls out of them without special syntax, as follows:

1. **Filter rows in SQL:**
   SELECT * FROM employees WHERE department = 'Engineering'
   In JQL:
   result = superstrict(employees, {(department, Engineering)})
2. **Project columns in SQL:**
   SELECT name, salary FROM employees
   In JQL:
   result = compose(D_{name, salary}, employees)
   projection is composition with a diagonal
3. **Join in SQL:**
   SELECT * FROM employees JOIN departments USING (dept_id)
   In JQL:
   result = employees ⋈ departments
   it matches on shared yin automatically
4. **Anti-join in SQL:**
   WHERE applicant_id NOT IN (SELECT applicant_id FROM employees)
   In JQL:
   result = substrict(applicants, employees)
   substriction is the algebraic anti-join
5. **Group and count in SQL**
   SELECT department, COUNT(*) FROM employees GROUP BY department
   In JQL:
   groups = partition_by(employees, department)
   yields a Horde: one Clan per department

Four things are worth drawing out, because they distinguish JQL's query model from SQL's at the root.

1. Filters are first-class Relations, not predicates — they can be saved and reused.
2. There are no NULLs: the absence of a Couplet means the fact does not exist, and three-valued logic never arises.

3. The schema is emergent — a Clan's "schema" is simply the union of the yins of its Relations, so Relations within one Clan may legitimately differ in shape.
4. And every operation is closed: it returns the same kind of structure it consumes, which is why queries chain indefinitely without the scaffolding of subqueries, Common Table Expressions (CTEs), or views.

This read algebra is deliberately side-effect-free.

Every JQL query is a pure function from structures to structures: it reads, and never writes.

That purity is the foundation on which the remaining four sublanguages are built — and it is what makes them so remarkably small.

2. Data Definition Language (DDL)

To define and change stored structure, JQL adds exactly one idea to the read algebra, and expresses everything else as sugar over it.

An open JabrDB database is a named Horde. We can say precisely that it is a namespace N mapping names to Clans, of which the Horde is the range.

A read query's leaf references resolve against N.

A definition, by contrast, does not transform a value passed into an expression — it transforms N itself.

So, the whole of definition and manipulation rests on two operations, and only these two carry side effects:

BIND n := <clan-expression> — evaluate the expression against the current state of N, then bind the result to the name n.

UNBIND n — remove the name n from N.

Everything below desugars to these. However many DDL statements the language offers on its surface — CREATE, DROP, RENAME, ALTER, and so on — each one is definable as some arrangement of those two primitives. They aren't independent operations with their own semantics; they're convenient names for particular BIND/UNBIND patterns.

Two features of this surface are not incidental; they are consequences of the read model, and they are why DDL in JQL is so slight.

1. The schema is never a separate object to alter. A column is not a declared thing — it is the set of Couplets, spread across a Clan's Relations, that share a yin. So DROP COLUMN is just a left-substriction lifted across the Clan, and RENAME COLUMN is just the read algebra's ρ made persistent. The emergent schema updates itself the moment the underlying Couplets change; there is no ALTER TABLE because there is no table-schema to alter. CREATE CLAN AS makes this vivid: its right-hand side is an ordinary read query, so defining a new Clan and querying an existing one are the same act, distinguished only by whether the result is given a name.

2. A drop never deletes a value. Every DROP in JQL removes Couplets — it is set-theoretic withdrawal, not destruction. The yin and yang elements involved remain members of the ground set G; only the association between them, the fact, is withdrawn. There are no NULLs and no tombstones, exactly as in the read algebra: the absence of a Couplet simply means the fact is no longer asserted. Dropping a Clan unbinds a name; dropping a column strips one yin from every Relation; in both cases the data persists as elements of G, and only the bindings or associations are removed.

3. Data Manipulation Language (DML)

If DDL acts on the names and columns of the Horde, DML acts on the Relations — the records — inside a Clan. It adds no further primitive: the three statements are again sugar over BIND and the read algebra.

1. CREATE RELATION IN <clan> { <yin>=<yang>, ... }
   or BIND clan := clan ∪ {R}

2. AMEND RELATION IN clan WHERE key APPEND { ... }
   [REPLACING]
   or BIND clan := unmatch(clan, key) ∪ amended(match(clan, key))

3. DROP RELATION IN <clan> (WHERE <key> | WHERE ANY)
   or BIND clan := unmatch(clan, key)

The one new concept DML needs is the key. Because a Clan is a set of Relations, a record is identified not by a row number but by content: a key K is itself a partial Relation — a handful of identifying Couplets such as {(emp_id, 1)} — and match(C, K) is the set of Relations in C that contain every Couplet of K.

Insertion (CREATE RELATION) is union with a literal Relation, and is idempotent: inserting a Relation already present is a harmless no-op, because sets do not hold duplicates.

Update (AMEND ... APPEND) merges new Couplets into every matched Relation by cross-union. Its one genuine decision concerns conflict: if the appended Couplets give a column a value the matched Relation already has under a different value, the default is union — the Relation simply becomes multi-valued on that column, which is the honest set-theoretic meaning of "append" — while the REPLACING modifier asks instead for a functional overwrite, stripping the old value before adding the new.

A Clan that declares a functional constraint on the column rejects a conflicting default-mode union and requires REPLACING. Deletion (DROP RELATION) is substriction: it removes the matched Relations and, as ever, deletes no value. Because an empty key would match — and therefore delete — every Relation, the empty case is guarded: emptying a Clan must be asked for explicitly with WHERE ANY.

It is worth noticing how soft the DDL/DML border has become. In SQL the two are firmly separated — one alters schema, the other alters data.

In JQL both are BINDs over the read algebra; CREATE CLAN AS is as much a query as a definition; and DROP COLUMN, nominally structural, is a Couplet-level manipulation that happens to change the emergent schema as a by-product.

The five-fold taxonomy is a familiar map, but in JQL definition and manipulation are the same country, because structure and data are made of the same algebraic stuff. One discipline is preserved deliberately, however: the payloads inside a manipulation statement — the AS query, the WHERE key, the APPEND literal — are full read JQL, so the whole power of the query language is available to compute what to write, but the act of writing is kept syntactically distinct. A query can never have a side effect, and a mutation can never be smuggled into an expression.

4. Data Control Language (DCL)

Here the SQL map and the algebra part company most sharply, and it is worth being plain about it. JQL, as specified for the desktop, has no

GRANT or REVOKE statements, and that absence is by design rather than omission.

Two reasons converge. The first is practical: JabrDB on the desktop is a single-user application in which a database is a folder on the user's own disk, so the meaningful access controls are the operating system's own file permissions, and a statement-level grant model would control nothing that is not already controlled. The second is deeper, and it is where the algebra reasserts itself. In this model, control is not a privilege attached to a table and held in a separate catalog; it is a property the data carries with it. A datum may be wrapped as a self-aware datum — a value carrying its own metadata facets, among them Ownership (whose data this is), Governance (what may be done with it, and by whom), and Access (the conditions under which it may be read or changed). Control, in other words, travels with the value rather than sitting beside it, and a mutation is gated by a governance check before its BIND is allowed to take effect.

This relocates the question that GRANT/REVOKE answers. Rather than asking "which users hold which privileges on this table," the algebraic model asks "what does this datum's governance facet permit, and does the requested operation satisfy it." Access-adjacent concerns that SQL bolts on as further features — column-level masking, federated identity — are likewise handled as companion surfaces over the same governance substrate rather than as inline language keywords.

A full statement-level control language — roles, grants, revocations expressed in JQL itself — is therefore a property of the multi-user server tier, not of the single-user desktop language defined here, and when it arrives it will rest on the governance facets just described rather than on a privilege catalog of its own. For the present chapter the honest summary is this: of the five sublanguages, control is the one that is, on the desktop, mostly latent — its substrate (the self-aware datum's governance) is real and carried by the data, but its surface syntax awaits the server. (A dedicated treatment of that surface is a natural companion to this chapter, and can be specified separately when the server tier is built.)

5. Transaction Control Language (TCL)

Transactions exist to make a group of changes all-or-nothing. JQL's read algebra needs none of this — a pure query alters nothing and so can never leave anything half-done — which means transaction control applies only to the mutation surface, and even there it is, today, almost entirely implicit.

On the desktop, every mutation statement is its own transaction. Each BIND or UNBIND is atomic over the namespace N: if evaluating the right-hand side fails — a bad reference, a constraint violation, a malformed literal — then N and the persistent store are left exactly as they were, and neither the in-memory Horde nor the on-disk store is touched. A successful statement updates both together; a failed one commits neither. Because JabrDB is single-writer by policy, there is no write contention and no locking model is required. The practical consequence is that the user already has the guarantee transactions are meant to provide — no statement can corrupt the store by half-completing — without writing a single transaction keyword.

Explicit, multi-statement transactions — grouping several mutations under one BEGIN … COMMIT, with ROLLBACK to abandon them — are deliberately deferred to the server tier. They are not needed on a single-user desktop, and when JabrDBServer introduces them they map directly onto transaction verbs the connectivity layer already declares (BeginTxn, CommitTxn, RollbackTxn), which until now have had nothing to wrap. So the transaction sublanguage, like the control sublanguage, is small on the desktop by design: per-statement atomicity now, explicit transaction blocks when more than one writer makes them meaningful.

The five sublanguages, redrawn

Laid over the algebraic tower, SQL's five-fold division comes out lopsided — and the lopsidedness is the point.

In JQL

- Query (DQL):This is almost the whole language – a closed algebra of perhaps a dozen operations, lifted across four levels, pure and freely composable.
- Definition (DDL): CREATE/DROP CLAN, DROP/RENAME COLUMN — it is sugar over one new primitive (BIND/UNBIND); thin, because the schema is emergent and never a separate object.
- Manipulation (DML): CREATE/AMEND/DROP RELATION. More sugar over the same primitive; its border with DDL is soft, because structure and data are one algebraic substance.

- Control (DCL): There is no inline GRANT/REVOKE; control is carried by the data as the self-aware datum's governance facets, with a statement surface reserved for the server.
- Transactions (TCL): Implicit per-statement atomicity on the desktop; explicit BEGIN/COMMIT/ROLLBACK reserved for the multi-writer server.

The shape of the language should be clear. JQL is, overwhelmingly, a query language — a pure read algebra that is closed and complete in its own right. Everything else is defined on top of it with a single addition: the power to bind a name in the Horde to a function of its current value. Definition and manipulation are that one idea dressed in convenient syntax; control is governance the data carries with it; transactions are the atomicity that one idea already guarantees, statement by statement. Where SQL draws five firm borders, JQL draws one — between reading and binding — and lets the algebra carry the rest. That is what it means to say JQL is not a language grafted onto a data model, but a consequence of one.

JQL v SQL

The SQL standard could be described as a partially successful sharing of query syntax so that there can be some data access capability that imperfectly spans multiple databases and data stores.

In the analysis that follows, we are going to highlight known problems and difficulties with SQL and explain, when it is the case, how data algebra with its query language JQL diminishes or resolves the problem.

The Original Sin: The Data Model

SQL is an offspring of the relational model of data, but for accuracy we will not consider the relational model itself, because SQL does not fully abide by the model. We'll take SQL as we find her.

The SQL data model is based on a single structural primitive: the table. In order to abide within this structure all data must be shoe-horned into rows and columns, regardless of its natural shape and its intended usage.

Every row in a table must conform to the same column set. This means that heterogeneous records don't fit. It's easy to think of examples of heterogenous record sets: product catalogs, medical records and so on. The RDBMS circumvents this problem using NULLS to indicate columns where a row has no value.. But that creates a thorny problem of itself, because the relational NULL has more than one meaning. It can mean:

- **Value unknown:** The data exists but we don't have it (a person's middle name we haven't collected yet).
- **Not applicable:** The attribute doesn't make sense for this entity (spouse name for an unmarried person).
- **Not yet assigned**: A value will exist in the future but doesn't yet (ship date for an unshipped order).
- **Explicitly empty:** Someone deliberately recorded "no value here" as a conscious data entry choice.
- **Default/uninitialized:** The column has no default and nothing was provided at insert time, so the system filled in NULL by fiat.
- **Deleted or redacted:** A value once existed but was removed.

So NULL actually means "I have no idea what this means." Two NULLS are not necessarily equal. This is the worst error of RDBMS and SQL and from it stem multiple sins.

But there's more. RDBMS do not allow nested structures. In practice that means that rows (or relations) cannot contain other rows. So hierarchical data or document-shaped data must be split up across multiple tables when stored and reassembled with joins when used. Put simply a RDBMS table cannot be used to store a sales order or an invoice.

This need to split up data objects into multiple tables led to the need to "normalize" data when splitting it up between tables. This becomes necessary in order to avoid data inconsistencies that can occur. So, normalization optimizes for data integrity and storage at the cost of query complexity — highly normalized schemas require more joins which are a performance overhead.

The need for normalization is a direct consequence of using the table as the single structural primitive.

RDBMS tables are not sets– they can best be described as bags. Duplicate rows are permitted unless explicitly constrained, which means that SQL does not implement set theory.

If a table does not have a primary key, its rows have no identity. There can be two identical rows that are indistinguishable, and yet both exist.

SQL has no native concept of the couplet and hence the meaning of data is inferred from the column in which a data item sits. The column-value association is implicit in the positional structure, but it is not explicit in the data model.

We can briefly summarize:

- Heterogeneous records do not fit in tables.
- The NULL value introduces and allows data value ambiguity.
- Nested structures do not fit in tables.
- Data normalization is required to ensure data integrity.
- Tables are not sets, they are bags which permit duplicate rows.
- In a table that has no primary key, there can be two rows that are identical.
- There is no explicit association between a data value and its meaning –no concept of couplet.

When we apply Data Algebra, all of these problems evaporate.

- Heterogeneous records fit happily into clans.
- There is no such thing as a NULL value.
- Nested structures fit naturally. The algebraic tower nests by construction and can conatin any nested structure.
- Data normalization is unnecessary.
- Tables are clans and they do not permit duplicate rows.
- Data Algebra does not need primary keys.
- The couplet is the basic atom of information in the algebraic tower.

SQL Composability Limitations

SQL has some awkward limitations. Chief among them is that SQL queries do not compose freely. You cannot easily feed the result set from one query into another. To do so, where it is even possible, means composing subqueries.

The output of a SQL SELECT is a result set, but feeding it into another operation requires subqueries, temporary tables, or views or CTEs (i.e., Common Table Expressions) which are temporary named result sets defined within a SQL statement using the WITH keyword. They exist only for the duration of that single query). In reality these are all work-arounds necessitated by the limitations of the language.

The major point is that in SQL there are no first-class intermediate results. You can't bind a query result to a name and reuse it naturally within the same statement without CTEs, which are syntactic sugar with scoping rules.

There are restrictions to subqueries too. Correlated subqueries re-execute per row. Subqueries in SELECT must return scalars. Subqueries in FROM require aliases. Each context imposes different constraints.

If you wanted to summarize all of these SQL composition problems you could say there is no query algebra in the mathematical sense. Within Data Algebra every operation consumes and produces the same kind of structure it receives. Compose, union, superstriction — all return Relations, Clans, or Hordes at the appropriate level. Operations simply chain when you need them to.

There are no subqueries. There are operations on structures. Superstriction, composition, and projection replace what SQL achieves through nested queries, but with no contextual restrictions.

In JQL, the algebraic query language, intermediate results are ordinary algebraic structures assigned to variables. Each variable holds a first-class Clan or Relation that can be used anywhere.

In Data Algebra, composition (∘) is a first-class operation. You can compose Relations, and by lifting, compose across the tower. The algebra is closed under all its operations. This is the core mathematical property that SQL lacks. There are no composability limitations.

SQL Type System Limitations

SQL does not standardize on types between different species of database. For example, VARCHAR, TEXT, NVARCHAR, CLOB, VARCHAR2 are all "strings" but they behave differently across vendors. Date/time types are notoriously incompatible.

In reality, SQL doesn't have a formal type system. In practice, data types are column attributes, not properties of the data itself.

Take these two things together and then consider SQL's CAST and CONVERT operations and you just know there are going to be problems. CAST (x AS type) is the actual SQL standard. Sadly, much was left to be "implementation-defined," so when you go across databases you can get wrong results. For example, Casting FLOAT to Decimal rounds differently between different species of database. CONVERT is non-standard. It is used by SQL Server and MySQL, but not by PostgreSQL. It suffers the same problems.

Polymorphism is awkward within SQL. This is the situation where you want to vary table names, column names, filter conditions or an entire query structure based on application logic. Static SQL cannot parameterize such things and thus such a capability is not possible unless you employ Dynamic SQL. Now all the major vendors support Dynamic SQL, but each with its own syntax so there is no portability across SQL dialects.

Common use-cases for Dynamic SQL include generic reporting tools, admin scripts, and pivot queries with variable columns. Aside from the dialect problem, the downside of Dynamic SQL is that it's harder to read and debug, and it bypasses compile-time syntax checking.

Data Algebra has none of these problems. All elements are drawn from a single universal ground set Γ, with a recursive type that includes atoms, integers, reals, and nested structures. There is one type system, not a vendor-specific zoo. The algebraic tower itself is the type system.

Operations are defined on structural properties, not on named schemas. For example, Superstriction works on any Clan that shares yin values with the filter Relation. If the column exists, the operation applies. No dynamic dispatch is needed.

The Expressive Power Limitations of SQL

1. Transitive closure

SQL has no native transitive closure. Graph traversal requires WITH RECURSIVE, which was added long after SQL was born and remains syntactically heavy and error-prone.

In data algebra, transitive closure is iterated composition: $R^+ = R \cup R \circ R \cup R \circ R \cup \ldots$ until a fixed point. This uses two existing operations (compose and union) — no new syntax required. Eliminated.

2. Pattern matching on sequences

SQL has no native pattern matching on sequences. SQL: SQL:2016 added MATCH_RECOGNIZE, but it's implemented by few vendors.

With data algebra, sequence pattern matching is not a native operation. However, sequences can be modeled as Relations with ordered yin values, and patterns can be expressed through composition and superstriction. This is an area where the algebra provides building blocks rather than a dedicated solution. This capability is partially addressed. It requires construction from primitives.

3. Group by problems

In SQL, GROUP BY collapses rows into aggregates. The original rows are gone. You cannot simultaneously access the group and its members without window functions (added in SQL:2003), so aggregation destroys structure. Grouping produces ephemeral aggregates, not a queryable structure of groups.

In data algebra, partition (grouping) produces a Horde, a set of Clans with each Clan being one group. The original Relations are preserved within each Clan. You can access both the group structure and the individual records simultaneously. Aggregation, if desired, is a separate operation applied to each Clan in the Horde. Problem resolved.

4. No set difference without column alignment

In SQL, EXCEPT requires both sides of the join to have the same column count and compatible types – a very limited kind of set difference.

In data algebra, the set difference operation (–) is defined on any two structures at the same level. Two Clans can be differenced regardless of their internal yin structure. The result contains all Relations present in the first but absent from the second. Problem resolved.

5. No native pivoting or unpivoting

In SQL, transposing rows to columns (PIVOT) requires vendor-specific syntax or dynamic SQL. The reverse of pivoting is equally awkward.

In data algebra, transpose (↔) is a first-class operation at every level of the tower. At the Relation level, it swaps yin and yang in every Couplet. At the Clan and Horde levels, it lifts. Pivoting is a natural algebraic operation, not a syntactic extension. Also, transpose is its own inverse. Unpivoting is just transposing back. Problem resolved.

6. Limited string processing

In SQL, regular expressions are vendor-specific. Recursive string parsing is impossible.

Data algebra operates on Elements, not strings. String processing at the character level is outside the scope of the algebraic operations per se. String functions would be part of the host language or a standard library, not the algebra itself. This is a shared limitation that data algebra has with SQL implementations.

7. Limited or no array/list operations

In SQL, arrays were added late (SQL:2003, SQL:2016) and support varies wildly across vendors.

In data algebra, ordered sequences are modeled as Relations with integer yin values (n-tuples): {(0,a), (1,b), (2,c)}. All standard operations apply to them. There is no separate array type with separate operations — sequences are Relations. Problem resolved.

SQL's Semantic Limitations

The following set of issues arises because of SQL's semantic limitations.

1. Column order with UNION

SQL: UNION matches columns by position, not by name. This is fragile and semantically wrong.

Data algebra's Union operates on set membership. Two Relations are unioned by combining their Couplets. Column names (yin values) are explicit in the Couplets, so matching is by name, not position. Problem resolved.

2. Implicit type coercion

SQL: Silent conversions (e.g., string to number) produce unexpected results rather than errors.

Data algebra's Elements have explicit types. An atom "42" and an integer 42 are different Elements. There is no implicit coercion — comparison is structural identity. Problem resolved.

3. Ambiguous column references

In SQL, without table qualification, column names in joins can be ambiguous, and resolution rules differ across vendors.

Data algebra has no implicit column references. Operations specify structures explicitly. Natural join matches on shared yin values, but those values are explicit in the Couplets. There is no ambiguity to resolve. Problem resolved.

4. SELECT * is fragile

In SQL, adding a column to a table silently changes every SELECT * query against it.

Data algebra has no equivalent of SELECT *. Operations are explicit: you operate on a Clan with composition, projection, or superstriction, each of which specifies what it acts on. Schema changes in the source Clan do not silently alter the operation's semantics — if a yin value is absent, it simply doesn't match. Problem resolved.

5. The evaluation order is not the syntactic order

In SQL, SELECT is written first but evaluated after FROM, WHERE, GROUP BY, and HAVING. This confuses every beginner and creates aliasing scope problems. HAVING exists only because of the evaluation order. It is a post-aggregation WHERE, necessitated by SQL's own syntax design.

In data algebra, the query language JQL is an expression language. Operations compose left-to-right or by assignment to variables. The evaluation order is the expression order. There is no disconnect between what you write and what executes. Problem eliminated.

SQL's Multi-Model Limitations

1. No native graph queries

In SQL, graph data requires recursive Common Table Expressions (CTEs) or vendor extensions (Oracle CONNECT BY, SQL Server's graph tables, etc.).

In data algebra, a directed graph is a Relation. Graph traversal is composition. Transitive closure is iterated composition to a fixed point. The four standard graph types (directed, simple, directed labeled, simple labeled) and their transformations are all expressible within the algebra. Eliminated.

2. No native document model

In SQL, JSON support (JSON_VALUE, JSON_QUERY, JSON_TABLE) was bolted on in SQL:2016 and varies dramatically across vendors.

With data algebra, a document is a nested structure of Couplets — a Relation whose yang values may themselves be Relations, Clans, or Hordes. The recursive Element type handles arbitrary nesting. Documents are not a separate model; they are algebraic structures. Eliminated.

3. No native time-series operations

In SQL, temporal queries require window functions, self-joins, and date arithmetic — all manual.

In data algebra, time-series data is naturally modeled as a Clan with a temporal yin (timestamp or interval as a yin value in each Relation). Comparison across time points is composition or superstriction against temporal filters. While the algebra doesn't provide dedicated time-series functions (moving averages, interpolation), the structural representation is clean and the basic operations apply naturally. Partially addressed — structure is native, but specialized temporal functions would need a standard library.

4. No native key-value model

In SQL, key-value access patterns require single-row lookups that bypass SQL's set-oriented design.

With data algebra, a key-value store is a yin-functional Relation: each yin maps to exactly one yang. Lookup is simply applying the Relation to a yin value. This is not a separate model — it is a property of a Relation. Problem eliminated by subsumption.

5. No native XML

In SQL, SQL/XML exists but is verbose and poorly supported outside DB2 and Oracle.

With data algebra, XML, like JSON, is a nested document structure. It maps naturally to nested Couplets and Relations. The same algebraic operations that handle documents handle XML. No separate language is needed. Eliminated by subsumption.

6. Separate language per model

In SQL, each new data shape (graph, document, spatial, temporal) spawns new syntax rather than being handled by existing operations.

In data algebra: One algebra handles all data shapes. The ~12 operations, lifted across four levels of the tower, cover tables, graphs, documents, key-value pairs, time series, and any other data shape representable as nested sets of ordered pairs. No syntactic extension is needed because the algebra is already complete. This is the central claim of the framework.

SQL Dialect Fragmentation

1. No true standard compliance

In SQL, no database implements the full SQL standard. Every vendor cherry-picks and extends.

Data algebra is mathematically defined. There is one specification: the algebraic tower with its operations. A compliant implementation either implements the operations correctly or it doesn't. There is no room for dialect variation because the operations are mathematical, not syntactic.

2. Pagination differs everywhere

In SQL, LIMIT (MySQL/PostgreSQL), TOP (SQL Server), FETCH FIRST (DB2/standard), ROWNUM (Oracle legacy). Pagination differs.

In data algebra, pagination is not a core algebraic operation — it is an interface concern. However, the algebra provides the structural tools: n-tuples with integer yin values allow positional access, and superstriction against a range Relation achieves slicing. The implementation is uniform because the underlying operation is. The problem is eliminated at the algebraic level; pagination UI remains an implementation concern.

3. String concatenation differs

In SQL string concatenation differs between vendors.

In data algebra, string concatenation is a scalar function, not an algebraic operation. It belongs to the host language or a standard function library, not the algebra itself. However, because there is one algebra and one expression language (JQL), there is one syntax for calling any such function.

4. Quoting identifiers differs

In SQL quoting identifiers differ. Double quotes, backticks, etc..

In data algebra, identifiers are Elements — atoms within the ground set. There is no quoting mechanism because identifiers are data, not syntax.

5. Date/time literals differ

In SQL formats, functions, interval arithmetic, and timezone handling are all vendor-specific.

In data algebra, date/time values are Elements in the ground set. Their representation and arithmetic are defined once in the Element type system, not per-vendor. Problem eliminated at the type level.

6. Auto-increment differs

In SQL vendor approaches to auto-increment vary.

With data algebra, auto-increment is a storage concern, not an algebraic operation. The algebra deals with sets of Couplets, not with how surrogate keys are generated. Key generation is delegated to the implementation layer. Not applicable — outside the algebra's scope.

7. Upsert/merge differs

In SQL upsert/metrge differs: MERGE (standard), ON CONFLICT (PostgreSQL), ON DUPLICATE KEY (MySQL), each with different semantics.

In data algebra: An upsert isa union with replacement semantics. If the Clan is yin-functional (each key maps to one record), union of the existing Clan with the new Relation either adds or replaces. The operation is simply union or a combination of substriction and union. One algebra, one semantics.

SQL Optimization and Execution Limitations

1. Optimizer opacity

In SQL, the query optimizer is a black box. Semantically equivalent queries can have vastly different execution plans with no way to predict which.

With data algebra, the algebra defines what an operation means, not how it executes. Optimization is still an implementation concern, and a JabrDB optimizer would face similar challenges in choosing execution strategies. However, the algebraic form is more amenable to algebraic rewriting rules (associativity, commutativity, distributivity of composition over union, etc.), which are provably correct transformations. The problem is partially addressed — algebraic identities provide a principled basis for optimization, but the optimizer is still an implementation challenge.

2. Hint syntax is vendor-specific

In SQL, when the optimizer fails, overriding it requires non-portable syntax.

With data algebra, if an implementation provides optimizer hints, they would be defined once for that implementation. Since there is one algebra, there is no multi-vendor hint syntax problem. However, whether hints are needed at all depends on the quality of the optimizer.

3. No cost model visibility

In SQL, the user cannot see or influence the cost model that drives optimization.

With data algebra, this is an implementation concern, not a language limitation. A JabrDB implementation could expose its cost model or not. The algebra itself is silent on execution costs. This is a limitation shared with SQL — it is an implementation design choice, not a language property.

4. Correlated subqueries are performance traps

In SQL, they look like simple filters but execute as nested loops.

In data algebra, there are no correlated subqueries. What SQL achieves through correlation, the algebra achieves through composition and superstriction — operations whose execution strategy is the implementation's responsibility, but whose semantics never imply "re-execute per row."

5. View performance is unpredictable

In SQL, views may be inlined, materialized, or neither, depending on the optimizer and the vendor.

With data algebra, a named algebraic expression is a definition, not a view. Whether it is evaluated eagerly or lazily is an implementation choice, but it is a single implementation making a single choice — not a per-vendor lottery. The implementation still makes choices, but there is no multi-vendor unpredictability.

SQL Security and Injection

1. SQL injection

In SQL, because queries are strings, interpolating user input creates one of the most exploited vulnerability classes in computing history, where the hacker adds SQL commands to an input string. Parameterized queries mitigate but don't eliminate this architectural flaw.

In data algebra, JQL expressions are algebraic structures, not text strings sent over a wire. User input is data (Elements), never operations. Injection is architecturally impossible because the boundary between code and data is enforced by the type system, not by escaping conventions. The problem is eliminated by the architecture.

2. No row-level security in the language

In SQL, row-level access control is a vendor extension, not a language feature.

With data algebra, row-level security is not a core algebraic operation. It would be implemented as superstriction against a permission Relation — structurally natural but not built into the algebra as a dedicated feature. A security layer would apply a filter Clan (representing the user's access rights) via superstriction before returning results. The problem is addressable within the algebra, but not a built-in feature.

SQL Metadata and Introspection

1. INFORMATION_SCHEMA varies

In SQL, the standard metadata schema is implemented inconsistently. System catalogs differ across vendors.

In data algebra, metadata is data. The catalog of a JabrDB instance is itself a Clan (or Horde) — queryable with the same operations as any other data.

There is no separate metadata language or schema. One algebra queries both data and metadata.

2. No self-describing results

In SQL, a query result doesn't carry its own schema in a machine-readable way without additional API calls.

With data algebra, a Clan is self-describing. Its yin (the union of all yin values of its Relations) is its schema. The schema travels with the data because it is part of the data. No separate API call is needed. Eliminated.

Philosophical Limitations

1. Language, not algebra

SQL is a natural-language-inspired syntax grafted onto a data model. It must grow syntactically to handle new requirements, because it has no algebraic closure.

Data aalgebra is, by definition, an algebra — a closed system of operations over a defined set of structures. JQL is a thin expression language over this algebra. New requirements do not require new syntax; they are handled by existing operations applied at the appropriate level of the tower. This is the foundational design principle.

2. Codd's model, incompletely

SQL doesn't faithfully implement the relational model. Codd himself criticized SQL for allowing duplicates, NULLs, and column ordering.

Data algebra subsumes and extends the relational model. It enforces set semantics (no duplicates), eliminates NULLs, and makes column association explicit through Couplets. It then goes beyond Codd by providing the tower (Clans and Hordes) and cross-operations that the original relational model lacked.

3. No universal data representation

SQL cannot represent its own queries as data within the same model (there is no homoiconicity).

In data algebra, an AET (Algebraic Expression Tree) is itself an algebraic structure — a tree of operations and operands that can be represented as nested Couplets and Relations. The algebra can represent its own expressions as data.

4. Vendor lock-in by design

In SQL, the dialect problem means SQL is portable in theory but not in practice, creating structural lock-in.

With data algebra: There is one algebra. A compliant implementation implements the defined operations over the four-level tower. Vendor lock-in through dialect variation is structurally impossible because there is nothing to vary — the operations are mathematical definitions, not syntactic conventions. Eliminated by architecture.

Summary

Our analysis of SQL identiified 56 different issues with SQL, many of which are caused by variances among dialects, and some of which are limitations. In only six instances does data algebra suffer from the same limitations, although it is no worse. The six partially-addressed/shared items are sequence pattern matching, string processing, time-series functions, optimizer opacity, cost-model visibility androw-level security.

Chapter 15: The Event Horizon

"An invasion of armies can be resisted, but not an idea whose time has come."

~ Victor Hugo

The Road From Here

We have spent considerable time developing JabrDB, and yet the honest truth is that we are at the beginning of a journey, not the end of one — a journey that may well last for decades. It is worth saying plainly what has been achieved and what has not. The mathematics is finished. The algebra of data is complete, closed, and proven; there is nothing left to discover in its foundations.

What is not finished — what has barely begun — is the building. A proof of concept is not an ecosystem, and a database that runs on a single desktop is not yet an answer to the problem of the data archipelago. This final chapter is about the distance between the two, and how we propose to cross it.

The distinction matters because the two kinds of work are not the same. The work behind us was mathematical and conceptual: showing that couplets, relations, clans and hordes form a single coherent structure into which all data fits. The work ahead is engineering. None of it requires a new theorem. All of it requires effort, many developers, conviction and commitment and of course, time. Much of what will follow is engineering that has not yet been funded.

One Engine, Many Personalities

JabrDB was never intended to be a single database. It was built as an algebraic engine that is capable of wearing many faces. Because the algebra subsumes every shape that data can take – the table, the document, the graph, the key-value pair, the multidimensional cube, the time series, the vector — a single engine built on the algebra can present itself as any of them.

We call these personalities. A relational personality offers tables and SQL to those who arrive from the traditional database world. A document personality treats each relation as a record with its own emergent document shape. A key-value personality reduces to point lookups. An OLAP personality presents dimensions, measures and cubes. A graph personality reads couplets as edges. A cartographic personality places data in space.

Nevertheless, each is a lens onto the same algebraic core, not a separate product re-implemented from scratch.

This is a genuinely different way to build data software, and it is one of the more consequential things the algebra makes possible. The industry has spent forty years producing a separate engine for every data shape — a relational database here, a document store there, a graph database, a vector database, a column store, each with its own query language, its own storage engine, its own operational discipline.

The algebra collapses that proliferation. The shapes were never fundamentally different; they were different surface presentations of the same underlying structure, and nobody had the mathematics to see it. A personality is a configuration of one engine, not a fresh act of engineering. The roster of personalities will grow over time, but the engine beneath them does not multiply with it.

The Gating Need: An SDK

If there is one thing that must be built before almost anything else, it is a software development kit.

The reason is simple, and the history of software bears it out repeatedly: adoption follows tools. A mathematical idea, however sound, spreads through an industry only when working developers can build on it without having to understand its foundations first. The relational model spread because developers got SQL and a library to call; they did not have to read Ted Codd's papers to use a database. Object orientation spread because languages and frameworks made it the path of least resistance. The algebra of data will spread the same way or not at all.

At present, building an algebraic application means, in effect, re-implementing the algebra. That is an absurd barrier and it must be implemented. What developers need is the algebra delivered as a library — the four levels of the tower, the full operator set, and the in-memory reference implementation, packaged so that any competent programmer can embed algebraic data structures and algebraic operations in their own software with a few lines of code.

The in-memory engine is, after all, already the reference implementation of the algebra itself; the task is to expose it cleanly, document it well, and make it available in the languages developers actually use.

An SDK changes the character of the whole enterprise. Until it exists, data algebra is something we build. Once it exists, data algebra is something anyone can build with — and that is the point at which an idea stops depending on its inventors and starts to propagate on its own. We regard the SDK as the single most urgent practical item on the road ahead. Everything else will be easier once it is in place.

The Student Dynamic

Currently, students are the most valuable audience data algebra can have, and they might be easy to attract. The advantage this technology has aside from the usefulness of the software, is that it is instructional - at least for students of mathematics and computer science. It could legitimately be included as part of a college course.

For this reason we have deliberately made the software inexpensive– a lifetime license for just $15.00. Students are the future workforce, and they are one of the three forces that will carry the data algebra into general use. If we capture a sufficient number of students, we will seed the next decade of data professionals. They also come unencumbered: a student has not yet spent fifteen years acquiring SQL habits that must be painfully unlearned, or set to one side. With a student, the algebraic data model meets an open mind rather than an entrenched one. And students talk. A cohort that learns on JabrDB carries it into internships, dissertations, and first jobs, doing the evangelism we could never afford to buy.

With this in mind our bundle almost designed itself. The book supplies the theory; JabrDB supplies the practice, with its built-in tutorials and its step-by-step decomposition of queries into algebra. Together they are a course in a box — and priced for a student, with a free or near-free tier, the barrier to trying it is negligible. Academic adoption carries a second benefit: universities offer no organized resistance to the paradigm, so they are the softest ground on which to plant it.

But this strategy makes two releases unavoidable. First, the tool must run on Windows and Linux. The overwhelming majority of students do not work on Macs, and a Mac-only product simply excludes the audience we most want; cross-platform is therefore a precondition, not an enhancement. Second, we must ship a few personalities as proof of concept — relational and graph at least — because these are exactly the database types students meet in their courses. Demonstrating that one algebraic engine can be all three, on the hardware they already own, proves the universality claim in the most persuasive possible way: in front of them, on their own screens.

An Honest Admission: We Have Only Built for the Desktop

We owe it to the reader to be candid about a limitation. Everything we have built so far — JabrDB and one or two other database personalities — runs only on the desktop; on a single machine, local storage and for a single user.

This was the right place to start, and we would make the same choice again. The desktop is where an idea can be made tangible. It is where a developer can drag a CSV file onto a window and watch it become a clan, type an algebraic expression and see the result, decompose a query step by step and finally understand what a join actually is. The desktop made the algebra visible and usable, and proving the model on real data, in a tool a person can hold in their hands, was a necessary first act.

But the desktop is not where the world's data lives. Data in the real world is shared, networked, concurrent, and large. It is accessed by many users at once, it must survive the machine that created it, and it rarely sits still. A database that cannot leave the desktop is little more than a demo as far as the algebra is concerned. And it plays no part in the data infrastructure that makes up the bulk of the world's vast archipelago of data.

A Spine for the Data: The Metadata Resource

Before data algebra can connect the archipelago, it needs a catalog — and the catalog must itself be algebraic.

One of the deepest claims of this book is that all data is already in valid algebraic form, but most of it is not in useful algebraic form, because the meaning of the data — its metadata — is locked away in the programs that use it rather than declared where others can find it. The remedy is a metadata resource: a catalog that records, for every clan and horde it knows about, the structure of the data, its origin, its lineage, and the rules that govern its use.

The elegant part is that this catalog need not be a thing apart. In the algebra, the catalog is simply more data — a clan whose relations each describe another clan, a clan of clans. It is queried with exactly the same operations used for the data it describes. There is no separate metadata language, no second system to learn and maintain.

"Show me every clan whose structure includes a timestamp" is an ordinary algebraic query against the catalog, not a special API. This is what we mean by catalog as data, and it is the spine on which everything more ambitious depends: federation, governance, and ultimately the global search capability we have dared to dream about.

From Desktop to Server

The decisive engineering leap is from the desktop application to the client-server system. The desktop apps we have built must become clients — clients to server versions of JabrDB and, in time, to other databases, algebraic or otherwise.

This heralds a move from a personal tool to a piece of infrastructure, and it brings with it everything that distinguishes a real database from a demonstration: concurrent access by many users, durable shared storage, transactional integrity under contention, and scale beyond a single machine's memory. The server holds the data; the desktop application becomes a window onto it. The same algebraic console, the same operator toolbar, the same step-by-step query decomposition that today work against local clans will work, unchanged from the user's point of view, against clans that live on a server and are shared across an organisation.

What travels between client and server is, naturally, the algebra. An algebraic expression is not a string of text to be parsed at the far end; it is a structured object — an expression tree of algebraic operations over named clans and hordes. The client constructs the tree, the wire protocol carries it, and the server evaluates it and returns algebraic structures in reply.

Because the expression tree is unambiguous, there is none of the dialect translation that bedevils conventional client-server databases. The operation means the same thing on both ends of the wire. We will not dwell on the protocol's internals here; it is enough to say that a frozen, well-defined wire protocol for carrying algebraic expressions already exists in specification, and that it is the channel through which the desktop becomes a client.

Reaching the World as It Is

Most of the world's data will never move into an algebraic database, and it does not need to. The final piece of the road is the means by which algebraic applications reach the data stores that already exist — relational databases, document stores, graph databases, file systems, and the rest.

This is the role of algebraic connectivity. We have it in mind to design and implement, what we call ADBC — Algebraic Database Connectivity. Its place in the algebraic ecosystem is fully analogous to the place ODBC occupies in the SQL world: a uniform interface through which an application reaches any data store, whatever its native format.

The difference is that where ODBC ships SQL strings and receives flat tables, ADBC ships algebraic expression trees and receives algebraic structures, with no loss of shape along the way. A document arrives as a relation, a table as a clan, a graph edge as a couplet; nothing is flattened to fit. Around ADBC will sit a family of connectors and gateways that translate between the algebra and the native protocols of the stores it reaches.

The consequence, taken to its conclusion, is that an algebraic application becomes a universal client to a heterogeneous world. The application speaks one language — the algebra — and the connectors handle the translation to whatever lies on the other side. The archipelago does not have to be rebuilt to become navigable; it has to be made algebraically accessible, one connector at a time, and the connectors are how the existing world is drawn into the algebraic fold without being disturbed.

Meeting Developers Where They Are

A final practical point. So far our tools have been native to one platform. If data algebra is to spread, its tools must reach developers wherever they work — on Windows and Linux as well as on the Mac, in the browser as well as on the desktop. The engine is the same everywhere because the mathematics is the same everywhere; what varies is only the shell around it. Broadening that reach is unglamorous work, but it is the kind of work that decides whether an idea stays a curiosity or becomes a commonplace.

The Work Is Engineering, Not Discovery

It is worth restating the through-line of this chapter, because it is easy to lose sight of amid the list of things to build. Not one item on this road requires new mathematics. The personalities, the SDK, the metadata resource, the server, the connectors, the cross-platform tools — every one of them is a matter of construction, not of theory. The hard intellectual work, the part that might have failed, is already done and cannot be undone. What remains is to build, and building is a question of will and effort and time, not of waiting for a discovery that may never come.

That is an unusually fortunate position for a young technology to be in. Most new ideas in software arrive half-formed, their foundations still shifting beneath them, and much of the early labor goes into finding out whether they work at all. Data algebra is past that point. Its foundations are settled. The risk ahead is not that the mathematics will fail; it is only that the building will go unfunded, unstaffed, or unattempted. That is a risk of resolve, not of soundness.

The Beginning, Restated

We opened this chapter by confessing that we are at the beginning of a journey, and it is fitting to end the book on the same note rather than on a flourish of completion. JabrDB and its relatives are the first steps, not the destination. They prove that the algebra works, that data really can be reunited under a single mathematical structure, and that the structure can be put to practical use by ordinary people working with ordinary data. That is a great deal to have proven. It is also only the start.

The data is still broken into billions of pieces, scattered across the digital archipelago, and the mathematics that can reunite those pieces now exists, is sound, and is waiting to be applied. The road from here is long, but for the first time it is a road and not a wilderness. It has a direction, a foundation, and a destination that can be described. What it needs now is travelers.

Appendices

Appendix A: Symbols and Notation

The table below lists all the mathematical symbols used in this book aside from the familiar algebraic symbols +, −, ×, ÷, =, !, <, and >.

Mathematical symbols	
Symbol	Description
{ }	Braces indicate sets
:=	Is defined as
∈	Belongs to
∉	Does not belong to
⊂	Subset of
⊊	Subset of and not equal to
| |	Cardinality
∘	Composition
↔	Transpose
x	Cartesian product
∅	The empty set
∪	Union
∩	Intersection
′	Complement
−	Difference
▼	Cross-union
▲	Cross-intersection
▷	Superstriction
◁	Substriction
▶	Cross-superstriction
◀	Cross-substriction
⋈	Natural join

Notation standards

The table below provides a reference for all typographic standards used throughout this book. They are used to distinguish different mathematical elements and structures from each other.

Notation standards		
Item	Convention	Example
element of a set	regular italic	a
couplet	regular italic with superscript	a^b
set	regular italic	A
set of sets	bold: font = Gauss	$\boldsymbol{\mathcal{A}}$
genesis set	regular: font = Greek	Γ
power set	regular	P
sets of numbers	regular: font = X Company	$\mathbb{N}$
relation	bold italic	$\boldsymbol{R}$
clan	regular: font = X Company	$\mathbb{C}$
function	bold italic	$\boldsymbol{f}$
partition	regular italic: font = Greek	$\mathit{\Pi}$

Bibliography

This Bibliography includes both books and papers that may interest readers who wish to acquire a deeper understanding of data algebra and related topics.

Currently the only further information on data algebra is the original data algebra book and a paper that was also used by the author in producing this book:

The Algebra of Data: A Foundation for the Data Economy by Sherman, G. J. and Bloor R.

"Data Algebra: Hiding in Plain Sight" (2012), by Sherman, G. J. and Underbrink, J. C.,

The following are recommended mathematical texts:

Halmos, P. R., *Naïve Set Theory* **(Van Nostrand, 1960).**

Suppes, P., *Axiomatic Set Theory* **(Dover, 1972).**

Dummit, D. S. and Foote, R. M., *Abstract Algebra* **(Wiley, 2003).**

The following are papers relevant to the contents of this book:

Kuratowski, K., "Sur la notion de l'ordre dans la théorie des ensembles," *Fundamenta Mathematicae,* **2 (1921): 129-131.**

Tarski, A. P., "On the calculus of relations," in *Journal of Symbolic Logic,* **6 (1941): 73-89.**

Codd, E. F., "A Relational Model of Data for Large Shared Data Banks," *Communications of the ACM* **13, 6 (1970): 377-387.**

Author's Biographical Details

Robin Bloor, Ph.D.

Robin Bloor was once a software developer and has recently taken up the profession again. Prior to that he spent most of his professional life as a technology analyst and IT consultant. He is also a frequent blogger and published author. He has a degree in Mathematics from Nottingham University (UK), a Ph.D. in Computer Science from Wolverhampton University (UK).

www.ingramcontent.com/pod-product-compliance
Lightning Source LLC
Chambersburg PA
CBHW071436230926
51846CB00001B/250
9781957278193